VISUAL SCIENCE
MECHANICS

John Freeman and Martin Hollins

Silver Burdett Company

Editor Margaret Conroy
Design Richard Garratt
Consultant Len Treharne
Picture Research Caroline Mitchell
Production Susan Mead

First published 1983
Macdonald & Co.
(Publishers) Ltd.
Maxwell House
74 Worship Street
London EC2A 2EN

© Macdonald & Co. 1983

Adapted and Published in
the United States by
Silver Burdett Company,
Morristown, New Jersey

1986 Printing

ISBN 0-382-06716-9 (Lib. Bdg.)
ISBN 0-382-09001-2

The Library of Congress has cataloged the
first printing of this work as follows:

Freeman, John.
    Mechanics
    (Visual science)
    Bibliography: p.
    Includes index.
    Summary: Discusses the phenomena of
force, movement, pressure, and friction,
the mechanics involved in moving things,
and the means by which machines work.
    1. Mechanics — Juvenile literature.   2.
Force and energy — Juvenile literature. 3.
Power (Mechanics) — Juvenile literature.
4. Power resources — Juvenile literature.
[1. Mechanics.   2. Force and energy.   3.
Power (Mechanics)]   I. Hollins, Martin.
II. Title.   III. Series.
QC127.4.F74   1983
620.1   83-50224
ISBN 0-382-06716-9

Cover: A floating crane.
Right: Motocross.

Contents

How do we move our bodies?

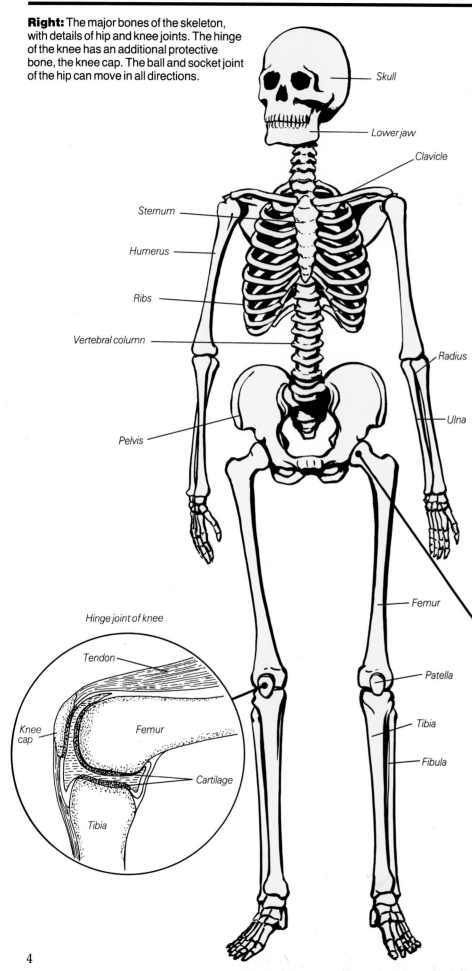

Right: The major bones of the skeleton, with details of hip and knee joints. The hinge of the knee has an additional protective bone, the knee cap. The ball and socket joint of the hip can move in all directions.

Skull

Lower jaw

Clavicle

Sternum

Humerus

Ribs

Vertebral column

Radius

Ulna

Pelvis

Femur

Patella

Tibia

Fibula

Hinge joint of knee

Tendon

Knee cap

Femur

Tibia

Cartilage

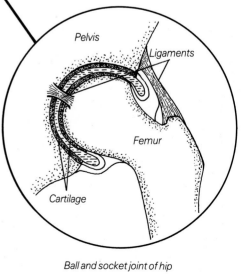

Pelvis

Ligaments

Femur

Cartilage

Ball and socket joint of hip

Our bodies can make a very large number of different movements. Some, like running and jumping, need lots of energy. Others need strength; for example, lifting and pulling heavy weights. In contrast, we can also make the very careful movements needed to write with a pen or to thread a needle. Often the most impressive movements that the human body can make are a combination of all the different types of movement, such as those of a gymnast.

How do people become such experts? It helps to be large and strong if you are a weight lifter, and to be light and supple if you are a gymnast. But these differences are mainly developed as a result of practice and training. Every body is built in the same way but each can be changed by the way it is used.

Muscles

All body movements are made by using muscles to pull parts of the body. The body contains about 200 bones, each one light but strong. Most important, the bones are rigid to give the body a solid frame – the skeleton. Bones are connected to each other by flexible joints so that they can move around. The bones are held together by string-like ligaments. Smooth cartilage covers the surfaces of the joints so the bones can move easily, and it also absorbs shocks.

Muscles are bundles of fibres which are attached to the bones by tendons. Muscles pull on bones by shortening their own length, that is, by contracting. They also become thicker and harder when this happens. You can feel this if

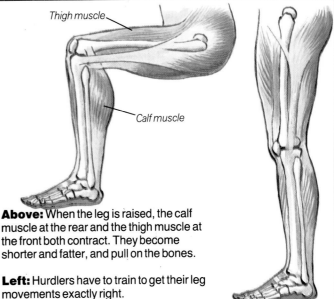

Above: When the leg is raised, the calf muscle at the rear and the thigh muscle at the front both contract. They become shorter and fatter, and pull on the bones.

Left: Hurdlers have to train to get their leg movements exactly right.

you hold the back of your leg as you lift your lower leg backwards.

Muscles are usually arranged in opposing pairs. One muscle pulls the leg up, the other pulls it down. It is possible to tense both opposing muscles together so that the bones are held steady – for example, when we want to hold something in position.

Muscle fibres are soft and stretchy. In the meat you eat you can see (animal) muscle fibres. They behave like elastic bands and cannot push, only pull. This is a tension force. Each muscle fibre contains a nerve which is connected to the brain, so that it can carry a message to tell the muscle how to move.

Left: Fine and detailed decoration of pottery requires great skill. Arm movements have to be carefully controlled.

Right: 1. Friction makes it hard to pull someone over a floor.
2. Water resistance slows down a swimmer.
3. Air resistance against the parachute reduces the effect of gravity as a person falls.

Energy

Using our muscles makes us tired. This is because we are using up energy. When a muscle is tensed, a chemical change occurs and we feel this as tiredness. While a muscle is resting, these chemicals are removed. To continue the movement requires more energy. Some movements continue all the time. The heart must beat and the lungs must take in air if we are to stay alive, so we need a continuous supply of energy.

Our bodies will only work at certain temperatures – we are 'warm-blooded' animals. We sometimes use up energy to keep warm. When we are working hard, the energy used also makes us warm and we sweat to get rid of the extra heat.

Our energy supply is the food we eat. Food supplies many things that the body requires: proteins, vitamins, minerals, water and roughage. We also need carbohydrates which give us most of the energy we use. These are contained in foods such as bread and rice.

Forces

Energy is used up because whenever we move something we do work against a force. Lifting a book involves moving

against the force of gravity. We jump upwards against gravity, and gravity pulls us back to the ground again. This force makes a body falling from a height move faster and faster, until it reaches a maximum speed which is too high for people to survive the impact of hitting the ground. A parachute makes the speed slow enough to land safely.

The extra resistance to falling which is provided by the parachute is due to the air it collects. Whenever we move through the air, it has to get out of our way, and this causes resistance. Water is heavier and has a lot more resistance or drag. Moving over the ground produces a resistance force.

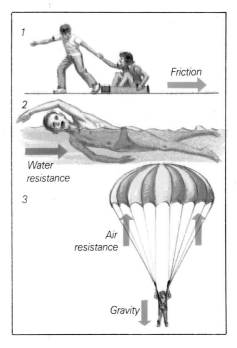

1
Friction

2
Water resistance

3
Air resistance

Gravity

Force and movement

A force may be simply described as a push or a pull. We can all sense this kind of force acting on our bodies, and we can push and pull things ourselves. When we do this, the objects – a dog on a lead or a stationary car, for example – often move. We cannot move a house this way, though! Of course, objects can fall without anyone 'forcing' them. So the connection between force and movement is not always obvious.

Gravity

In the seventeenth century, two famous scientists made discoveries about gravity. Galileo Galilei is remembered for a demonstration in which he dropped balls from the leaning tower of Pisa in Italy, and Isaac Newton is associated with the story about an apple falling on his head. Galileo found that all objects fall at the same rate (provided that there is not much air resistance). Newton showed that this is due to the force of gravity. The size of an object does not matter because gravity exerts the same force on all objects. It makes falling objects increase speed, or accelerate.

The acceleration due to gravity on Earth is always about 10 metres per second per second. This means that every second, the speed of a falling object increases by 10 metres per second. A greater force causes a greater acceleration. The powerful engine of the Renault Fuego GTX provides enough force to accelerate the car from 1.33 metres per second to 2.33 metres per second in 9 seconds, whereas the less powerful Renault 4 takes 15.5 seconds to do this.

Force is also needed to slow things down. But forces which change the motion of an object are unbalanced forces. If something is not moving, the forces on it must be balanced, as shown in the tug-of-war.

Units of measurement

Forces can be measured by a spring balance. The unit of force is the newton (symbol N). One newton is the force which will cause a mass of one kilogram to increase its speed by one metre per second every second. By a coincidence, the force of gravity on an apple is about one newton!

When you lift an object, you are using a force to do work. We can calculate the amount of work done from the simple formula:

work done = force × distance.

If the force is measured in newtons and the distance is measured in metres, then

Above: The trampolinist uses the resistance of the trampoline to stop her downward fall but the elasticity of the trampoline then provides her upward acceleration.

Above: The pull of both teams on the rope causes a tension force in the rope. This force is the same on each team. It is balanced, for each person, by the force of the ground on his or her feet. To win, one team must create a bigger force at the ground by digging their heels in harder.

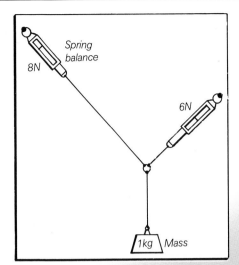

Spring balance

8N

6N

1kg Mass

Equal forces

Left: The force of gravity on the 1 kg mass is 10N downwards. It is supported by two spring balances. These have forces to the left and upwards of 8N; to the right and upwards of 6N. These two forces are combining to give a total upward force of 10N which balances the hanging 1kg mass.

Direction

To know what effect a force will have, it is necessary to know its direction as well as its size. Forces are called vectors for this reason. The force of gravity on a falling apple is one newton downwards, which is why it falls downwards. If it is supported by your hand, you will be providing an upward, balancing, force of one newton. If instead you push it downwards with a force of one newton, it will accelerate downwards faster than when it is only affected by gravity. The total downward force is now 2 newtons.

If an object is supported by forces at angles to it, these must be combined in a special way, using what is called the triangle of forces.

Below: The two tugs are pulling the tanker forwards. Each is also pulling sideways, but in opposite directions, so these forces balance out.

the work done is measured in joules (symbol J). (This is named after another famous scientist.) For example, lifting that apple would need a force of one newton, so in raising it 2 metres, 2 joules of work are done.

Energy is the capacity to do work and it is measured in the same units. Lifting the apple takes 2 joules of energy.

Power is the rate at which work is done. It is measured in joules per second, or watts (symbol W). If the apple was lifted in 2 seconds, the power would be 2/2, which equals one watt.

The force of gravity on the apple is its weight. This is different from the mass of the apple which is measured in kilograms. The mass of the apple is always the same but on the Moon the apple would weigh less because gravity is less. Weight is measured in newtons.

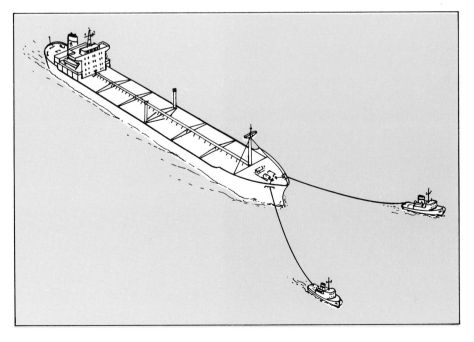

The limits of human endeavour

The human body is like a machine that is capable of an amazing variety of tasks. These may need the careful co-ordination of several parts of the body at once, or lots of energy, or both at the same time. You may remember your first attempts to swim or cycle or play a ball game. Until the machine – your body – worked properly, you sank, fell, or missed the ball!

How the body moves has been described. The engine that drives the 'machine' is the heart. It pumps the blood, which contains oxygen, around the body. This oxygen enables fuel from the food we eat to release the energy we need. The heart's pumping action is called a beat. If the action stops for more than a few minutes, the person will die. How long is your heart beat? Over a lifetime of 80 years, the heart will beat 3000 million times. That is nearly equal to the number of people in the world! It is not very powerful though – with an output of about 2.25 watts, it is similar to an electric motor in a toy.

Energy needs

The body itself is, of course, much more powerful. The record in weight-lifting (on the back) is about 700 kilograms. That is the weight of about 10 people. Athletes can jump higher than the tallest person, and further than five people laid end to end. But human efforts do not go very far in satisfying our needs today. A person working hard all day would do about three million joules of work. We use this amount of energy in a small electric heater in one hour. One kilowatt hour (the energy consumed in supplying 1000 watts for one hour) of electrical energy costs about 5p and is almost equal to 3.5 million joules. We therefore have to use other sources of energy to meet all our needs. These include the fuels coal, oil and wood which can be burnt; nuclear fuels; solar power; and the movement of air and water. The energy is then used to drive machines to do the work we require.

Human adaptability

Our actions are controlled by our brains. Some, such as breathing and digesting food, continue without us having to think about them. Others, like swimming and cycling, we can learn so well that we don't have to think about them. The human brain weighs about one kilogram, and is a larger part of the body than the brain of any other living thing. It is made up of 30,000 million cells called neurones. Learning involves making connections between all these neurones. The vast number of possible connections explains why we can learn to remember and do so many things. Computers are still very simple compared to the human brain.

The body is also very adaptable to different conditions. People live all over this planet, in temperatures which can range from −88°C to +58°C. But people's bodies do not reach these temperatures – if the body temperature falls below 36°C we die of cold, and above 40.5°C we die of overheating. The body itself has ways of adjusting its temperature. Shivering warms us up because it is muscular movement, and blood does not flow so near the surface of the skin, so heat is kept in the body. If we get too hot the body sweats and the blood flow is sent to the skin, so we lose heat. But in extreme temperatures we have to assist our bodies by wearing suitable clothing.

Warm clothing can be furs or padded coats or woven wool. Wool traps air close to the body and this helps insulate it from the cold conditions. It is also necessary, in arctic conditions, to wear a waterproof and windproof outer layer.

Good insulators like wool are also useful to help keep the body cool. It is often worn by firemen and furnace operators. Covering the wool is a layer which will not burn. Asbestos has been used for the last 100 years. Nowadays, material such as glass fibre, carbon fibre and special plastics are more common.

Left: Russian super-heavyweight Vasili Alexeyev. The best weight-lifters in this class can lift over 250 kg, overhead.

Left: Jim Hines of the USA was the Olympic gold medallist in the 100 m in 1968. His record time was 9.95 seconds.

Below: Total protection for an Alaskan oil-pipeline worker who wears wool and padding for insulation and a windproof and waterproof outer covering. Even the face is completely covered.

Below: Undersea salvage divers. Rubber suits give protection against the cold. The air supply flows from the back tanks into the face mask.

Levers

Any mechanical arrangement which makes work easier for people is a machine. A simple machine does not use any fuel as energy source, nor does it reduce the amount of work to be done. It normally reduces the amount of force needed – the effort.

The lever is a simple machine which was discovered about 6000 years ago. It is very useful since a lever can make it much easier to lift heavy weights. Archimedes, a Greek philosopher, said that he could move the world if he had a long enough lever.'

The shadoof, a water lifting machine invented hundreds of years ago, is still in use in simple irrigation systems near the River Nile in Egypt. A shadoof consists of an upright pole with a horizontal pole fixed across the top on a pivot.

Attached at one end of the horizontal pole is a bucket on a rope and at the other is a counterbalancing weight. Because of the weight, it is easy to swing up a bucketful of water from the river. The bucket can then be swung on its rope so that the water can be tipped into an irrigation ditch.

Levers have different forms but all have a rigid arm, a pivot (or fulcrum), a place at which the load is positioned, and a place where the effort is applied.

The theory of levers

Many levers are designed to increase the force available. Examples are the crowbar used to move large rocks, the wheelbarrow and the spade. The wheelbarrow enables one person to lift and move heavy loads. The load is pos-

Above: Very large loads can be moved by using levers. On Easter Island, people used a lever to slowly force upright a stone statue weighing many tonnes.

itioned between the effort and the pivot (the wheel). The wheel allows some of the weight of the load to be carried by the ground without much friction.

Other levers are designed to increase the distance a force moves or the speed at which it moves. Examples are the human arm, a fishing rod, and a ballista. A ballista was a weapon used by the Romans for throwing rocks. A large, heavy rock was dropped on the effort pan to throw a smaller rock. A ballista could throw 10 kilogram rocks 100 metres.

Can you spot the difference between the 'force-increasing' levers and the 'dis-

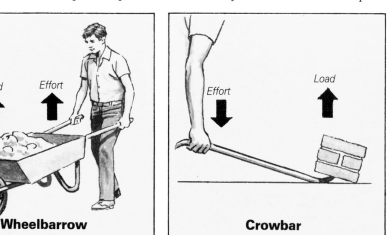

Wheelbarrow

Load   Effort

Crowbar

Effort   Load

Effort

Load

Spade

tance-increasing' levers? All the force-increasing levers have the load closer than the effort to the pivot. (Load-pivot distance is less than effort-pivot distance.) All the distance-increasing levers have the effort closer than the load to the pivot. (Effort-pivot distance is less than load-pivot distance.) Levers can have the pivot, load, and effort arranged in many ways and the lever arm can be many shapes as well, but one of these two rules always applies.

The effort needed to move a load with a particular lever can be worked out using this equation:

$$\text{effort needed} = \frac{\text{load-pivot distance}}{\text{effort-pivot distance}} \times \text{load}.$$

The effort is measured in newtons. The equation shows that for a given load, the effort can be reduced by increasing the effort-pivot distance. This means, for example, that using a longer spanner needs less effort than using a short one.

Classes of lever

Levers may sometimes be divided into three orders or classes. The first order of levers is the most common and the pivot is located between the effort and the load – for example, the crowbar. A second order lever has the load between the effort and the pivot, as in the wheelbarrow. The effort is applied between the load and the pivot in a third order lever, as in a fishing rod or the arm. All three types work on the same principles and they are only distinguished by the relative positions of the load, the effort and the pivot.

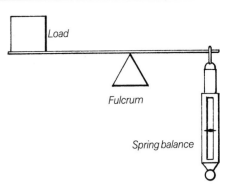

Above: Using a ruler and a spring balance, this apparatus can measure the effects of different loads and different fulcrum positions.

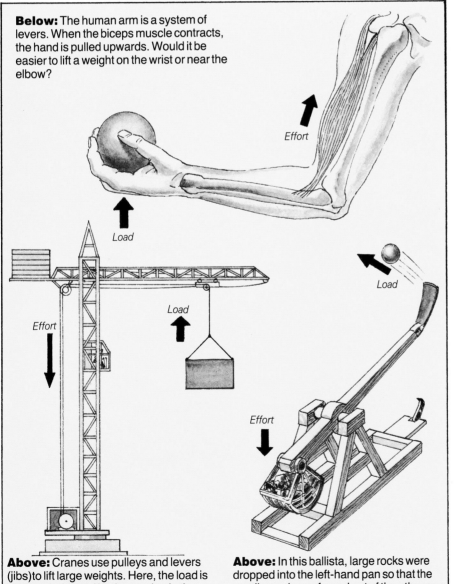

Below: The human arm is a system of levers. When the biceps muscle contracts, the hand is pulled upwards. Would it be easier to lift a weight on the wrist or near the elbow?

Above: Cranes use pulleys and levers (jibs) to lift large weights. Here, the load is counter-balanced by the tension in the cable. The counter-weight balances the jib weight so that the crane does not fall over.

Above: In this ballista, large rocks were dropped into the left-hand pan so that the smaller rock was forced out of the other pan, very fast. Notice that the pivot is not in the middle of the arm.

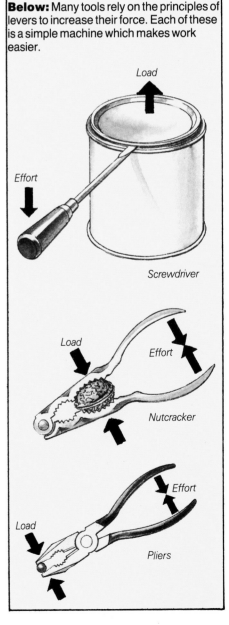

Below: Many tools rely on the principles of levers to increase their force. Each of these is a simple machine which makes work easier.

Wedges and screws

Like the lever, the wedge is a simple machine that was discovered thousands of years ago. A wedge is usually a block with triangular sides and it can be forced into a gap narrow-end first. The axe heads that stone-age people made out of flint were wedge-shaped.

Below: Splitting logs is made easier by using either a wedge-shaped axe, or by pushing in separate wedges. In both cases, the downward force is directed into a sideways force to push the wood apart.

A wedge increases the applied force (because it is concentrated at the thin edge), but decreases the distance through which it moves since the force pushes out to the sides as the wedge enters an object. A wedge can do the same job as a force-increasing lever, although it works in a different way. A thin wedge will increase the force more than a thicker wedge. Wedges are all around us; for example, axes, chisels, bolt cutters, knives, door stops.

Wedges can be used in building construction. If you look at an old building with a timber frame roof, you will probably be able to see wedges inserted into the joints. These were hammered in to take up the slack in the joint, and make it completely rigid.

Inclined planes

An inclined plane is like a wedge being used to lift something, but the object moves instead of the wedge. The enormous stone blocks that were used to build the Egyptian pyramids were probably slid into position up a long inclined plane of sand and earth that was later removed. Builders today use inclined planes or ramps. The shallower the slope of the inclined plane the easier it is to lift a load. However, friction makes very long inclined planes inefficient.

You will find inclined planes in other places; for instance, multi-storey car parks and where railway lines go up hills.

Left: Moving an armchair. The chair is heavy and difficult for one person to move. In order to lift furniture into a van, an inclined plane or ramp is often used.

Screws

Screws are among the most important of simple machines. Surprisingly, they are closely related to inclined planes. To see how, take a rectangular piece of paper about 15 centimetres by 30 centimetres and cut it along a diagonal. This triangle could be the side of an inclined plane. Now wrap it round a pencil, rolling from the 30 centimetre edge of the triangle, to obtain a shape like a screw.

As a screw is turned, the edge of the screw (the 'thread' of the screw) moves along fairly slowly – more slowly than the head of the screw turns. You should remember that a machine that decreases the distance a load moves, increases the force. Thus, a woodscrew holds two pieces of wood together more securely than a nail. The thread of the screw causes greater friction than a smooth nail.

A bolt is like a woodscrew that does not taper towards the end. Nuts fit on to the ends of bolts and as they engage on the thread, they move slowly along the bolt as it rotates. Either the nut or the bolt can be turned. When the force-increasing property of a bolt and nut is combined with a special force-increasing lever, the spanner, the resulting force can be very large indeed. The wheels on cars are held on securely in this way.

Bolts, nuts and woodscrews are among the most common examples of simple machines. You should be able to find at the very least 1000 in the average home!

Jacks

Jacks for lifting heavy weights in a straight line often use screw threads. Examples are the car jack, and the scaffold jack which is often to be seen on building sites. A common type of car jack consists of a vertical rotating bolt with a fixed nut joined by a bracket to the car which is to be lifted. As the bolt turns, the nut and bracket move up or down. Another type is the scissor jack. As the bolt is turned it forces the two opposite corners of the jack inwards, and this causes the top of the jack to rise.

The scaffold jack usually has a fixed bolt, and a rotating nut on which an outer tube rests. The nut is turned to raise or lower the tube. In order to obtain the very large forces needed to raise a scaffolding structure, a long lever arm is used to operate the scaffold jack. It can support large weights, such as those involved when a hole is knocked in a wall.

Above: A scissor jack used for lifting a car. As the screw is turned, the two halves of the jack are forced together, and the top is raised.

Below: A scaffold jack. The nut is turned to raise the top part while the bolt rests on the ground. A bar is often used as a lever to turn the nut.

Wheels and pulleys

Below: When the Ancient Egyptians built the pyramids, thousands of huge stone blocks were dragged up inclined planes of sand and earth. Rollers were used to make the task easier.

Even using an inclined plane, it is very difficult to move heavy weights uphill. The Egyptians building the pyramids discovered how to do this more easily, by using rollers. The rollers were simply wooden logs, smoothed to make them roll along. The very large stone blocks used to build the pyramids were supported on many such log rollers.

Development of the wheel

Eventually it was discovered that, for carrying lighter loads, the mass of the log rollers could be reduced by slicing them into thin circles and using two of these wheels, connected by an axle which was attached to the load. Four wheels joined in pairs by two axles were used to make a cart. Still later, it was discovered that the wheel did not have to be solid, and also that it could be made out of metal instead of wood. This is the key to much modern engineering; few machines are without wheels or rollers of some type.

Almost all vehicles run on wheels: cars, lorries, bicycles and trains, for example. Bulldozers run on wheels inside tracks. Even vehicles that do not run on wheels almost always have wheels somewhere in their machinery.

Using pulleys

A wheel with a groove, in the outer edge, for carrying a string or rope is called a pulley. Pulleys can change both the direction of a force and its position of application.

When builders are working on a tall building they often fix a pulley at the top of the building so that a worker at the bottom can attach a load to the rope and then pull it up. In a single pulley system, the effort force and the load force are about the same. When several pulleys are used at once, the effort needed is reduced. In order to work out the approximate effort needed to lift a load, count the number of strings holding up the load and divide the load force by this

number to find the effort needed. (Really, there is only one rope in most cases – look at the diagrams!)

Pulley blocks with several sets of pulleys were often used (and still are) on sailing ships. This was so that fewer sailors were needed to change the set of the sail.

Differential pulleys

A differential pulley system allows very

Right: All wheels have a characteristic circular shape, so that they roll easily. The cart wheel, the car wheel and the train wheel all developed from the log roller.

large loads to be lifted without a large number of pulleys. Instead, two pulleys of slightly different sizes are connected together. This system can generate very large forces. Car engine hoists work like this. These enable one person to lift easily and safely a car engine which has a mass of over 200 kilograms.

Pulleys can also be connected by continuous ropes or belts to provide a drive at different speeds or in different directions. You can find pulleys like this in car engines (the fan belt), upright vacuum cleaners, and washing machines. In general, when a large pulley drives a second smaller pulley, the second pulley will rotate faster. A small pulley will drive a second larger pulley more slowly, the speeds in proportion as the ratio of the sizes of the pulleys. For instance, if a rotating 10 centimetre diameter pulley drives a 20 centimetre diameter pulley, the large pulley will rotate at half the speed of the smaller pulley. If the pulleys were changed round the driven 10 centimetre diameter pulley would rotate at twice the speed of the larger pulley.

In an automatic washing machine, the tub rotates much more slowly than the motor during washing; to do this the motor drives a small pulley, which drives a large pulley connected to the tub. When the washing machine is spinning the wet clothes, the tub has to rotate much faster even though the motor always rotates at the same speed. In order to achieve this higher spin speed, the tub may then be driven by a pulley that is much smaller.

'DAF' cars use pulleys of variable sizes instead of a gearbox. As the car speeds up, instead of different gears being selected, the pulleys are squeezed in or out to change their effective size. This method can give a smooth drive. Unfortunately, it can also lead to a lot of wear on the pulley belt.

Right: The simple pulley (left) just changes the direction of the pull from down to up. But the four-string pulley also greatly reduces the effort needed. The reduction of effort is related to the number of strings. Here the effort is about a quarter of the load force.

Left: This 25cm-wide belt connects two pulleys in the mechanism of a pulverising mill. This provides a simple, compact and efficient drive for the machine.

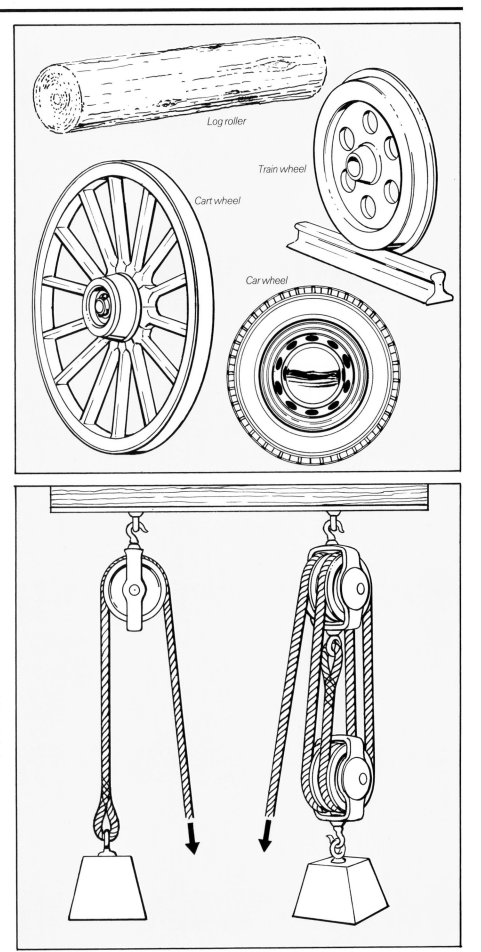

Log roller

Train wheel

Cart wheel

Car wheel

Gears

Above: A rack-and-pinion gear is used to stop mountain trains slipping when the slope of the rails is so great that the wheels cannot otherwise grip the rails.

Below: The differential gearbox transmits the drive from the car engine to the drive wheels using crown gears. When the car travels round corners, the differential engages to make the wheels rotate at different speeds. The outer wheel travels faster round the corner.

Above: Two spur gears meshed together. The teeth are carefully shaped to give a smooth transmission of power with little friction.

may slip. Because gears have to be manufactured very precisely, they are more expensive than pulleys, and if a pulley will work well enough, it is usually used instead.

To summarize, gears are used rather than pulleys either when it is important that there is no slip at all (for instance in clocks) or when large forces are transmitted (for instance in a car gearbox).

Rates of rotation

The relative rates of rotation of a pair of gears can be worked out easily from the relative numbers of teeth on each gear. When the number of teeth on the driven gear increases, the rate of rotation decreases, and vice versa. To work out the actual speeds, first count the teeth on each gear, then use this formula:

$$\frac{\text{output}}{\text{speed}} = \frac{\text{input speed} \times \text{input teeth}}{\text{output teeth}}$$

Suppose a gear with 32 teeth is turning a gear with 48 teeth and is rotating at 3 revolutions per second. The rotation rate of the driven gear is given by:

$$\frac{\text{output}}{\text{speed}} = \frac{3 \times 32}{48} = \frac{2 \text{ revolutions per}}{\text{second}}.$$

Gear systems

The gears considered so far are simple spur gears but gears come in a great variety. For example, the crown gear changes the direction of rotation by 90°. Some gears are not connected directly, but by a chain. Bicycles usually use chains and gears to connect the pedals to the back wheel. Another type of gear is the rack-and-pinion, used to give good grip on the mountain railways and in car steering systems. Differential gears are fitted to cars to enable the driving wheels to rotate at different speeds when travelling round corners.

The worm gear, which is basically a

Gears are usually used to change the speed or direction of movement of rotating shafts.

Why not pulleys?

Gears are like pulleys in that their motion is usually circular and continuous, unlike the motion of levers and wedges. The advantage of using a gear rather than a pulley is that a gear cannot slip. The rope round a pulley, however,

Driving straight

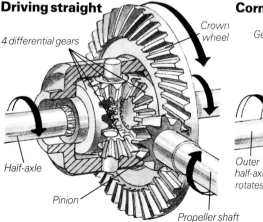

4 differential gears

Crown wheel

Half-axle

Pinion

Propeller shaft

Cornering

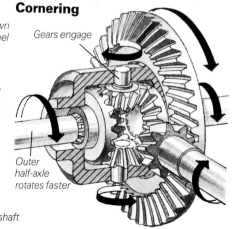

Gears engage

Outer half-axle rotates faster

Right: Power is transmitted from the bicycle pedals by the chain and two gears. The gear fixed to the rear wheel is small, and therefore rotates faster than the pedal gear, reducing the effort needed to make the wheels turn.

Centre right: This is a five-speed Derailleur gear set. When the idler gear is moved, the chain is guided over smaller gears.

screw thread engaged with a spur gear, causes the spur to move by one tooth for a complete rotation of the worm. This gives a very low gear ratio. Helical gears are used in car gearboxes in preference to spur gears because they are much quieter. Since helical gears are more expensive, car makers often economize by using a simple spur gear for the reverse, which is why cars often sound 'whiny' when reversing.

In all these cases, the method for calculating speeds is the same, based on the number of teeth.

Derailleur gears
Cars and bicycles are equipped with different gear ratios for different conditions. This is because the engine (or a person's muscles!) works best within a certain range of speeds.

The gearbox of a car is very complicated, but the principle is the same as in a derailleur five-speed gear on a bicycle. In this gear set, the particular gear the chain passes over is determined by the chain guide. This is moved by a cable which is controlled by a lever on the frame of the cycle.

At low speeds, a low gear ratio (large gear) is chosen. The pedals have to turn a lot to move the bicycle far but good acceleration is obtained. As the speed increases, higher gear ratios (smaller gears) are selected so that the pedals do not have to turn as fast. The available acceleration is decreased. At top speed, fifth gear is engaged. The pedals move much more slowly than they would in first gear. When an uphill slope is reached, a lower gear has to be selected to enable the cyclist to continue, pedalling faster to overcome gravity and friction forces and get up the hill.

Right: As the car steering wheel is turned, the rack is moved along inside the outer casing. This causes both wheels to pivot. The rack and casing are sealed by rubber 'boots' to avoid corrosion.

Sprocket has fewer teeth

Chain wheel has many teeth

Pedals travel through a smaller distance than bicycle wheels.

Connecting chain

Low gear

High gear

Gear change mechanism

Idler gear

Pinion

Rack

Friction

When two surfaces rub against each other, a force called friction opposes the movement. This is because even smooth surfaces are actually rough when looked at under a microscope. This movement of the tiny ridges and pits against each other causes the friction force.

The amount of friction depends on the actual surfaces that are rubbing together: when rubber rubs on concrete, the friction is high; if the two surfaces are ice and steel, the friction is very low.

It is often necessary to control friction. High friction may be required, for instance in shoes and boots. If there were not a lot of friction between the soles of shoes and the ground, a person walking would just slide about instead of pushing forwards. This is what happens to you on an icy surface. Sand is often thrown on to icy pavements and roads to increase the friction. (Salt is also used; this helps to melt the ice.) The friction in car brakes stops the car. In this case a special composite material is made to rub against a steel surface in a drum or disc brake.

Reducing friction

There are three ways to reduce friction between two surfaces. One is to lubricate the surfaces by putting a thin layer of liquid between the surfaces so that they do not touch each other. Oil is a common lubricant that is used in car engines, bicycle bearings and door hinges. Oil is not the only possible lubricant; even water will do. For instance, one of the reasons why ice skates are so effective at reducing friction is that a thin film of water forms between the blade and the ice. On a wet road, water can build up between the car tyres and the road. This dangerous lubrication is called 'aqua-planing', and makes the steering and brakes completely ineffective.

Gases have an even lower friction than liquids. Hovercraft use this fact by supporting themselves on a cushion of air.

A second way to reduce friction is to choose the surfaces carefully. PTFE (poly-tetrafluoroethylene), is a plastic used to coat non-stick cooking pans. PTFE causes very little friction with most other materials. (So does a banana skin, which is why it is so dangerous to tread on one!)

The third way to reduce friction is to change sliding friction to rolling friction, which is smaller. (This is why the wheel was such an important invention.) To see how small rolling friction is, take a

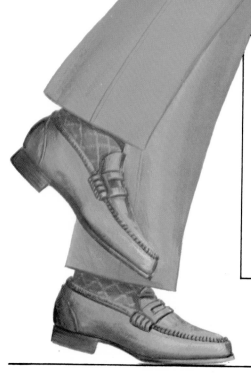

Right: People rely on friction between the ground and their shoes when walking. The leg muscles are trying to push the ground backwards so the body is pushed forwards.

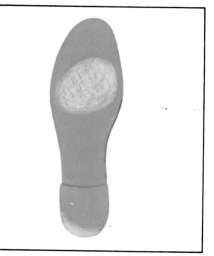

Below: The worn areas on a shoe are the parts which push against the ground. The wear is due to friction.

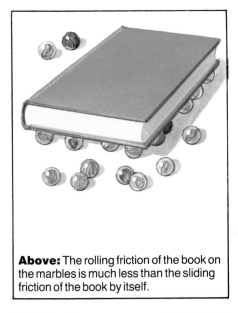

Above: The rolling friction of the book on the marbles is much less than the sliding friction of the book by itself.

Right: A skater can move much faster than a running person, partly because the friction between ice and steel is so low.

book and push it across a table. Now rest the book on a few round pencils and repeat. The book should have moved much more easily. If you now replace the pencils with marbles the book will be able to move freely in all directions. You have just used roller bearings and ball bearings. Almost all modern machines include sets of ball or roller bearings made of steel and lubricated with oil or grease. It is important to keep these bearings lubricated with oil. This not only reduces the friction but also stops the ball bearings rusting, which would prevent them from rotating freely.

Effects of friction

When friction and sliding both occur, there are two consequences that have not yet been mentioned. First, the energy of movement of the moving body (kinetic energy) is converted into heat. This is why you can burn your hands when you slide down a rope, and why brakes become hot after use. Car brakes can become very hot indeed, even red-hot. This can cause 'brake fade' since the high friction pads are less effective at high temperatures.

The second consequence of friction is wear. This will always occur whenever there is sliding. Brake pads are designed so that the brake pad will wear away much faster than the steel parts. This is because the brake pads are cheaper and easier to replace than the steel parts.

Above: The ball bearing in this electric drill allows it to rotate freely. Thus the work that the motor does turns the drill bit and is not wasted in friction.

The bicycle

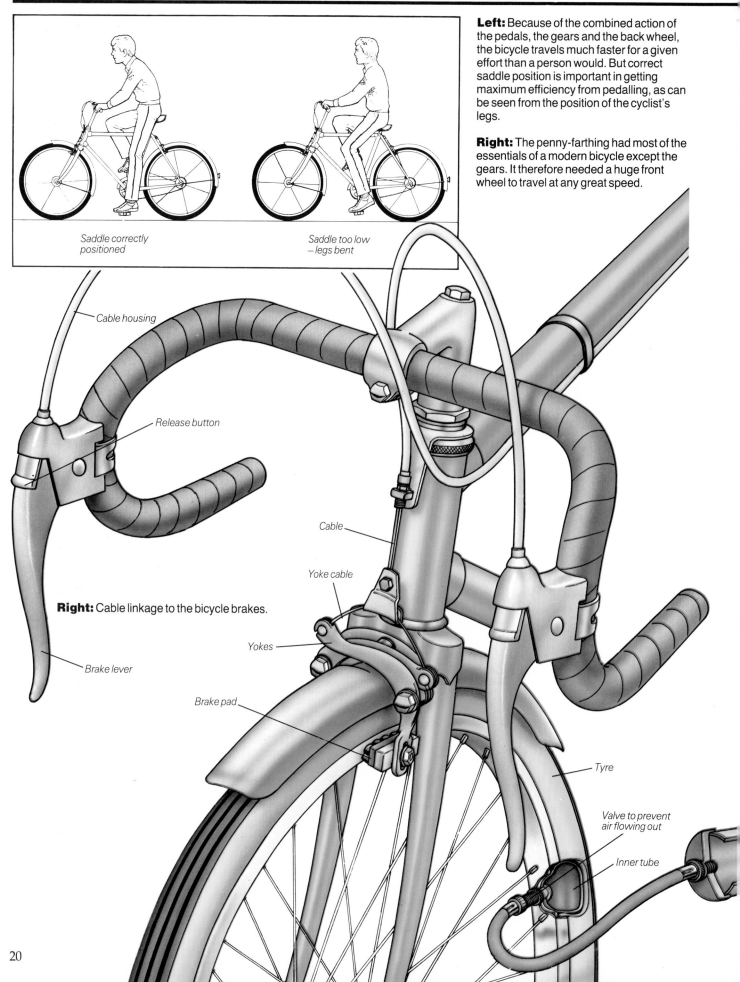

Left: Because of the combined action of the pedals, the gears and the back wheel, the bicycle travels much faster for a given effort than a person would. But correct saddle position is important in getting maximum efficiency from pedalling, as can be seen from the position of the cyclist's legs.

Right: The penny-farthing had most of the essentials of a modern bicycle except the gears. It therefore needed a huge front wheel to travel at any great speed.

Saddle correctly positioned

Saddle too low – legs bent

Cable housing

Release button

Cable

Yoke cable

Right: Cable linkage to the bicycle brakes.

Yokes

Brake lever

Brake pad

Tyre

Valve to prevent air flowing out

Inner tube

The bicycle uses many of the principles already explained to make personal transport easier and faster than walking or running. Riding a bicycle and running both utilize the same source of energy; that is, human leg muscles. How do bicycles make the same source of energy produce a greater effect?

Gears, bearings and tyres

An important energy saving results from the ability of the bicycle to move fast even though the legs are moving quite slowly. This is achieved by means of the pedals, which are levers turning the gears, which in turn transmit the power to the back wheel by the chain. Finally the rear wheel turns, and friction between the tyre and the road accelerates the bicycle.

This combination of simple components results in the feet moving much more slowly in a circle than the bicycle moves along the road. However, as has already been explained, a machine that increases the distance a load moves requires a greater effort. This is fine on the flat or downhill, as bicycles are designed to roll easily and freely. But when a bicycle is being propelled uphill, or needs to accelerate rapidly, the gearing needs to be changed to enable more force to be applied. Usually, a larger gear is selected at the back wheel. This results in the back wheel turning more slowly but with more force.

There are two types of gearing in common use, the three-speed enclosed gear and the five-speed derailleur gear. Some bicycles are fitted with a second gear at the pedals. This provides a total of 10 gear ratios and allows the cyclist to choose the best gear ratio for a very wide range of conditions.

You may have seen an old penny-farthing bicycle. The front wheel was made large because at that time no-one had thought of using gears, and the large wheel was the only available means of travelling fast. Of course, a penny-farthing has no free wheel, and the cyclist's feet had to be taken off the pedals when travelling downhill at speed!

Bicycles are designed to reduce friction to the absolute minimum. The wheels replace sliding friction with rolling friction, and they are fitted with ball bearings. Ball bearings are also used in the pedal crank and for the handlebars.

Racing cycles exaggerate several features present in ordinary bicycles. They are fitted with very hard and narrow tyres to reduce the rolling resistance to the absolute minimum; the cyclist takes up a higher, crouching posture that makes maximum use of leg muscles while reducing air resistance; and racing cycles are much lighter than ordinary bicycles.

Braking

Once a cycle has started moving, the cyclist has to consider the potential problem of stopping. Most cycles are equipped with friction pads which are pulled against the rim of the rotating wheel either by a cable or by a rod linkage. One disadvantage of this system is that the rim of the wheel easily becomes wet and 'lubricates' the brakes, with unfortunate and possibly dangerous effects! This can increase the stopping distance as much as nine-fold. While modern composite friction materials may help, it is better to be safe than sorry, and allow a longer distance to stop in wet weather.

Stability

People have often puzzled over the question, 'Why is a bicycle stable?', that is 'Why doesn't it fall over?'. In fact, the explanation is quite simple. If a bicycle is moving forwards and starts to fall to the left, the pivot at the handlebars will cause the wheel to turn left. This left hand turn tends to push the bicycle upright again. And of course, the opposite happens when the bicycle falls to the right. Cyclists very soon learn to apply just the right amount of tilt to compensate for any amount of turning force.

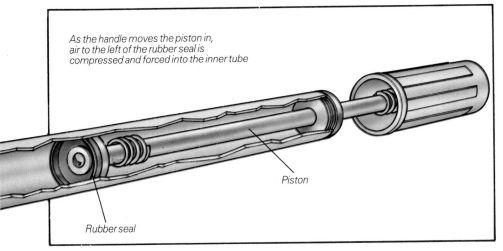

As the handle moves the piston in, air to the left of the rubber seal is compressed and forced into the inner tube

Piston

Rubber seal

Left: The pump is an essential part of any cyclist's kit. It enables the user to compress air to the high pressures needed to inflate the pneumatic tyres.

The car

The motor car, which was developed in the early years of this century, must be one of the most popular machines ever. It is a combination of many mechanisms so the workings of the car are best understood by dividing it into parts.

The engine

The engine is the most important part. This is where the power is developed to drive the car. The source of energy is usually petrol or diesel but engines have been developed to run on gas, or alcohol made from wood. The energy is released by burning the fuel. This has to be done in a controlled way to ensure maximum efficiency.

The petrol is pumped from the tank to the carburettor where it is sprayed

Four-stroke petrol engine

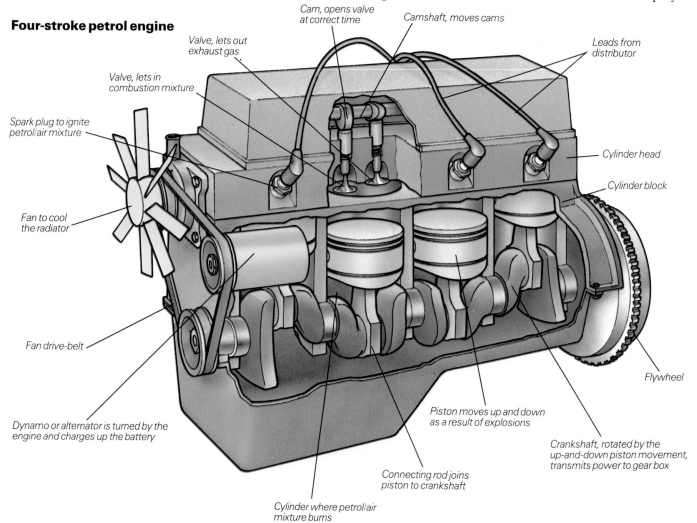

Cam, opens valve at correct time

Camshaft, moves cams

Valve, lets out exhaust gas

Valve, lets in combustion mixture

Leads from distributor

Spark plug to ignite petrol/air mixture

Cylinder head

Cylinder block

Fan to cool the radiator

Fan drive-belt

Flywheel

Dynamo or alternator is turned by the engine and charges up the battery

Piston moves up and down as a result of explosions

Crankshaft, rotated by the up-and-down piston movement, transmits power to gear box

Connecting rod joins piston to crankshaft

Cylinder where petrol/air mixture burns

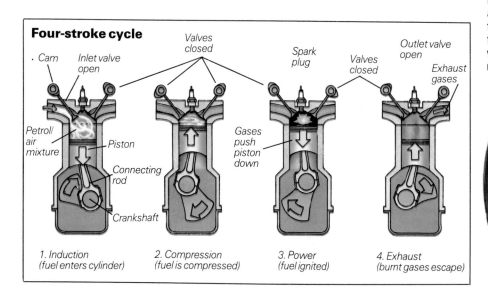

Four-stroke cycle

Cam

Inlet valve open

Valves closed

Spark plug

Valves closed

Outlet valve open

Exhaust gases

Petrol/air mixture

Piston

Connecting rod

Gases push piston down

Crankshaft

1. Induction (fuel enters cylinder)
2. Compression (fuel is compressed)
3. Power (fuel ignited)
4. Exhaust (burnt gases escape)

Below: Racing car tyres for dry weather are wide and smooth to give a good grip on the road. Ordinary car tyres have tread so that they grip the road well in any weather. In wet weather water is channelled away by means of the tread, so friction is maintained.

through a fine jet to make a vapour which is mixed with air. It is then sucked into the cylinder, or combustion chamber, where it is exploded by an electric spark from the spark plug. The force of the explosion is carefully controlled, in a sequence, by the camshaft, cams and valves to provide the smoothest possible flow of power. When an engine is cold, extra petrol can be added to the petrol/air mixture by using the choke control. More power is obtained by allowing more mixture into the engine by opening the throttle tube in the carburettor wider, which is done by pressing down the accelerator pedal.

Transmission

The crankshaft transmits a rotary motion from the engine, through the gearbox, to the wheels. This is called the transmission system. At the same time a belt drive from the pulley on the front of the shaft turns a fan to cool the engine and turns the alternator or generator coil to recharge the battery.

The engine is connected to the transmission system through the clutch. This is operated by a foot pedal (or automatically). It consists of three plates which can be separated to disconnect the gearbox when changing gear. If this were not

done, gear wheels would hit each other at a speed which could damage the teeth. The gearbox reduces the rotation of the crankshaft which is usually between 2000 and 5000 revolutions per minute, to the wheel rotation required to drive at the desired speed.

Cars usually have four gears which allow them to travel forwards at speeds of up to about 150 kilometres per hour, and one reverse gear. The correct gear is selected using the gear lever. The lowest or first gear reduces the rotation most, to enable the wheels to turn slowest. They, then have most power, so low gears are used on hills.

The propeller shaft runs from the gearbox to the differential and by means of its crown wheel drive, rotation passes to the axle. The crown wheel drive 'bends' the drive by 90°, and the differential allows the half shafts and wheels to rotate at different speeds. This is necessary to turn corners, when the outer wheel has to travel further.

Motion is then transmitted through the wheels and tyres to 'push the ground backwards'. Good grip is needed for this and it is provided by the rubber outer casing which has regular grooves, called the tread. If the tread is worn away, a car can skid on the road.

Making the car move

This whole sequence of events is operated by only three controls: the accelerator, clutch and gear selector. To control the car's movement requires several other systems. Direction is controlled by the steering wheel: stopping is controlled by the foot and hand brakes. The electrical system includes many indicators, as well as providing the ignition spark. The suspension system provides a comfortable ride. All of these systems have to work with little effort from the driver, to make a car convenient and safe to drive.

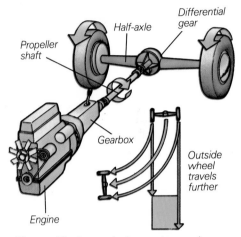

Above: The transmission system and (inset) the effect of the differential on the car wheels as it turns a corner.

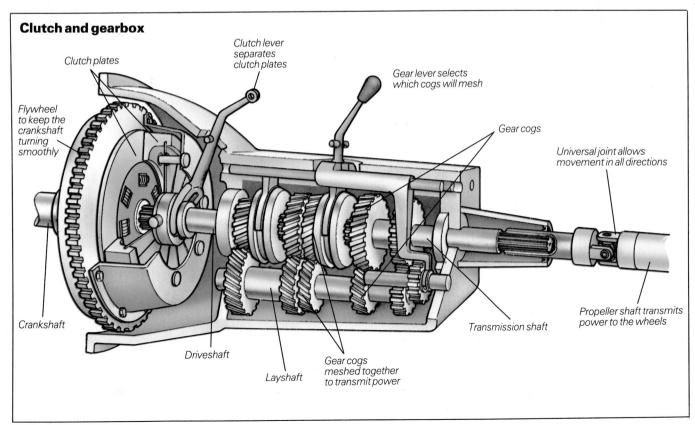

Clutch and gearbox

Clutch plates

Clutch lever separates clutch plates

Gear lever selects which cogs will mesh

Flywheel to keep the crankshaft turning smoothly

Gear cogs

Universal joint allows movement in all directions

Crankshaft

Driveshaft

Layshaft

Gear cogs meshed together to transmit power

Transmission shaft

Propeller shaft transmits power to the wheels

Safe driving

Fully protected passenger compartment

Well-padded steering column will collapse in a collision

Strengthened seats

Rear window electrically heated to keep clear

Seat belts for all passengers

Fuel tank well protected

Child-proof locks on rear doors

Locks will not burst open in a collision

Strong windscreen with good all-round vision

Steel tubes inside the doors to protect from a side collision

Two braking systems in case one fails

Headlight wipers and washers

It is about 70 years since cars became generally available to people, with the mass production of the famous Model T designed by Henry Ford of the USA. Although today's cars look very different, the basic principles on which they work are the same. The changes in appearance over the years have been mainly due to improvements in the comfort, convenience and simplicity of driving, which all contribute to greater safety. People do not have to be experts to drive cars. Most people soon pass their driving test, but many are later involved in accidents in which people are hurt and killed. How can this be avoided?

Preventing accidents

One way is for drivers to improve their driving skills. Another way is to have laws to try to stop bad, dangerous driving. And yet another way is to make sure that cars are safe to be on the road.

In most western countries, cars have to pass tests which check that they are working properly. The following points are usually tested:

✳ The steering system and transmission system must drive the car, in the right direction, perfectly.

✳ The suspension system must work properly so that the driver or car is not shaken about or loses a good hold on the road.

✳ All the brakes must work efficiently and evenly to stop the car quickly, without swerving.

✳ The wheels must run smoothly and the tyres must be the right type, with enough tread to grip.

✳ All the lights must work, so that other people can see the car and tell if it is going to turn, stop, or reverse. The headlamps must point in the right direction so the driver can see ahead but not dazzle other drivers.

✳ The horn, the windscreen wipers and washers must work.

✳ The exhaust system which carries away the hot burnt gases goes rusty quickly – it must be replaced when damaged.

✳ The whole of the car bodywork goes rusty in time. If this becomes too weak the car could just break apart, so the bodywork must be in a good state.

In Britain, it is illegal to drive a car more than 3 years old without it having passed such a test.

Safety features

Manufacturers are now introducing more safety features into car design. Often this adds to the expense, but makes the car more likely to be damaged rather than the people inside. Trials are still being done to find the best means of making cars safe.

The seat-belt is, so far, the only compulsory safety addition to cars. Its purpose is to stop the passengers when the car stops suddenly. If we were travelling in a car at 100 kilometres per hour and the car stopped suddenly we would continue to travel (through the windscreen) at this speed. This is because our bodies' inertia would keep us travelling forwards. A strong force must be applied to stop us, but not too strong or it would just injure us instead. That is why seat-belts have to be carefully designed and properly fitted. They stop the body going forward and head rests, or head restraints, stop the head going violently backwards which can damage or even break the neck.

If a passenger is thrown about in a car during an accident, injury can occur if

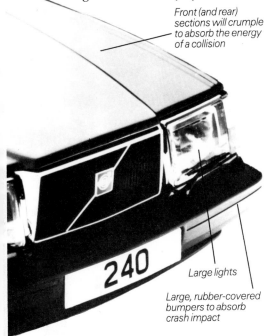

Front (and rear) sections will crumple to absorb the energy of a collision

Large lights

Large, rubber-covered bumpers to absorb crash impact

there are any sharp objects sticking out inside the car. Switches, door catches, ash trays and car instruments can all be designed to avoid causing injury. The steering column is a hazard to the driver if it is pushed in towards the driver's body, particularly if the wheel snaps off. The column can be made with extra joints so that instead it collapses easily. The car body can be constructed to improve safety in a collision. The passengers need to be seated within a rigid box, but the rest of the car can be allowed to crumple up to absorb the impact. More damage is done to the car – but less to the people inside.

Car tyres are a very important part of safe driving. Apart from having tread, they must be correctly inflated to give road-holding. Punctures can always occur and these can make the car go out of control if the tyre deflates quickly. Tyres have been developed that can avoid this. In one type a liquid is released

Above: Ford 'Model T' and modern (right-hand drive) cars. Design changes for increased safety include greater protection for the driver and better road-holding and stability due to the lower, more streamlined shape.

within the punctured tyre, sealing the hole and giving off vapour which re-inflates the tyre.

Future developments
Despite the dangers of accidents, car driving is so popular that cars cause

other problems too. Two are very obvious in large towns; traffic jams and exhaust fumes. In the fumes are poisons such as carbon monoxide and lead compounds. Yet another problem is that world supplies of oil will one day run out and so alternatives to the petrol engine are being researched. One possibility is the electric powered car.

Below: Seat-belts are rigorously tested by crashing cars with dummies in the driver's and passenger's seats.

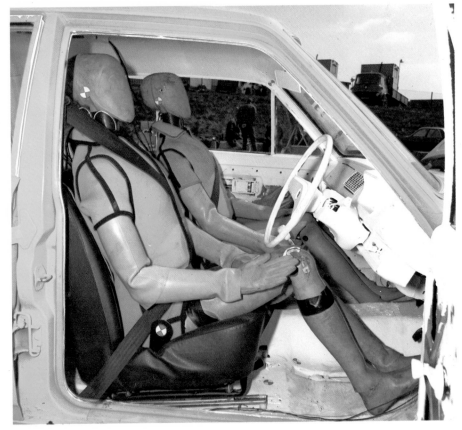

Trains

As a method of transportation trains have both advantages and disadvantages when compared with cars, lorries and aeroplanes. They are able to carry large loads fast and easily. However, because they need special track which is expensive to install and maintain, trains can only have a limited number of 'pick-up' points.

Trains can carry heavy loads because of their combination of strong steel wheels and rails. Unfortunately, although this combination gives extremely low rolling resistance, it also ensures low friction between the wheels and the rails. This means that the wheels slip if a train accelerates or slows down too quickly. In autumn, modern electric trains are often slowed down by the lubricating effect of dead leaves on the rails. Even in normal conditions, the distance taken to stop a train is much longer than is needed by a lorry or car. This is why the advance signalling system used on railways is so important.

Sources of power

Trains were almost all propelled by steam locomotives until well after 1950. Steam engines burn coal, and the heat is used to boil water to produce steam at high pressure. The steam enters a cylin-der where it forces a piston to move backwards and forwards, so turning the wheels.

Today, there are few steam locomotives working in Europe. One reason is that they are inefficient; only five per cent of the energy in the coal goes towards moving the train. Secondly, they produce a great deal of smoke and pollution. Thirdly, they require a great deal of maintenance; several hours for each day's work. Fourthly, they are very heavy and cause a lot of track wear.

Modern locomotives are of three main types: electric; diesel-electric; or diesel-mechanical. Electric locomotives are the simplest. They pick up electrical power, either from an overhead cable using a pantograph or from a third rail below them. The electric power is used to turn electric motors fixed to the axles of the trains. Diesel-electric locomotives use powerful diesel engines to turn electric generators which supply current to electric motors on the axles. The British High Speed Train is of this type. Diesel-mechanical locomotives use diesel engines connected to the wheels through gearboxes similar to those used by lorries. Many two-coach trains are diesel-mechanical. A train like this makes a 'changing-gear' noise, like lorries, as the

Above: The French TGV (*Train à très grand Vitesse*) is a 'super-train' which uses either high voltage overhead cables or gas turbine engines as power sources.

Below: In this cut-away diagram of an HST you can see the diesel engines and the generators which they power, and the axle-mounted traction motors.

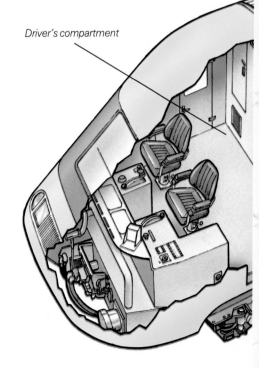

Driver's compartment

engine starts to turn the wheels.

Some countries have developed super-trains. These need special tracks because they travel extremely fast. Super-trains are always streamlined to minimize air resistance. The Japanese 'Bullet train' is powered by overhead cables and travels at up to 255 kilometres per hour.

All of these trains need special track which is less curved than usual, so that they do not have to slow down for the

Above: A modern electronic signal-box. The signals and points are operated electrically. The display shows the positions of the trains at all times.

Above: A mechanical signal-box. The signals and points are operated by the large levers which move rods and cables. An operator is needed to control them.

Guard's compartment

12-cylinder diesel engine drives generator

Generator provides power for electric motors

Electric motors drive wheels

bends. The British 'Advanced Passenger Train' (APT) is designed to use existing track by tilting into bends, in the same way as bicycles do.

Signalling
Railways of every sort need a complex system of signalling to avoid crashes. The arrangement used is called 'block working'; that is, only one train at a time is allowed in a section of track, called a block. The signals are designed to be 'fail-safe', which means that if there is an electrical or mechanical defect anywhere in the system the signals are set automatically at the 'stop' rather than 'all-clear'. In the same way, points have to be set correctly before a signal allows a train to enter them.

Ships

It was only about 500 years ago that ship design became more scientific. Before that ships had many sails and several masts, rising many metres above the decks, with the decks also built up high above the water level. The whole appearance may have been attractive but the ships could be 'top-heavy', especially when loaded with soldiers, cannons and supplies, and so were liable to capsize.

A famous capsize in English history is that of the *Mary Rose*, Henry the Eighth's flagship, which, when fully laden for battle, turned over in Portsmouth Harbour in 1545. About 700 sailors drowned because the ship sank immediately.

Modern ship design

Modern ships are very different. Their shape is designed to make them stable, they are built from different materials and they use different sources of power.

In the nineteenth century, sailing ships were gradually replaced by steamships. A famous early steamship was the *Great Britain*, designed by Isambard Kingdom Brunel and built in 1845. It was made of iron and powered by a screw propeller. It marked the beginning of modern ship design.

The engines of early steamships used reciprocating pistons, like those in the car engine but driven by steam. By the end of the nineteenth century, turbines had been found to be much more efficient.

This century the main change in ships has been in their source of power. Oil replaced coal at about the time of the first world war. It is much more convenient to use in ships of all sizes because it takes up less storage space and is lighter than coal. Nowadays, oil supplies are more expensive so other fuels are being tried. A few nuclear powered vessels have been built – the first was the American submarine Nautilus in 1954. Nuclear powered engines are very expensive. They need to be large with plenty of shielding for safety. They can go a long time without needing to refuel, so they are most useful in large submarines.

Buoyancy

It is easy to accept that wooden ships float. But how can steel ships stay up when a piece of steel sinks? How are ships designed so that they are unlikely to capsize and how do sailing ships manage to sail towards or into the wind which is acting against them?

When something floats, the force of

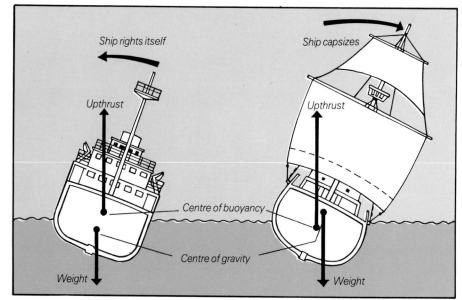

Above: The centre of gravity is the centre of the ship's weight. The upthrust force acts upwards through the centre of buoyancy. If the ship is top-heavy, it is likely to capsize.

Below: The force of the water on the keel balances the sideways force of the wind and prevents capsizing. In boats with small keels, sailors lean out on the windward side to help the balance.

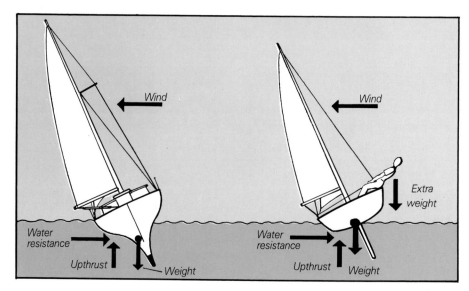

its weight is balanced by the upward force of the water. This force is called the upthrust. The size of the upthrust force depends on the weight of the water that is pushed out of the way by the object. This is called the displacement. A floating body sinks until the weight of the water displaced, and hence the upthrust, is equal to its weight.

A solid block of steel will not displace enough water to support it. If it is made into a hollow shape with air inside, then its underwater volume is greater and can displace more water. It will then sink only if it leaks and the air space becomes filled with water. This is why steel ships are able to float.

Sailing into the wind

No ship can sail straight into the wind but it is possible to sail close to it. The sail is set carefully at the correct angle to the wind so that it still catches it. Some force will then push the boat sideways, but some will also push it forwards, because of the force on the sails and the position of the boat's steering rudder. The sailors then have to compensate for the sideways movement so that they keep on the course they want. This is why boats have to sail on a zig-zag course when sailors want to go straight into the wind.

Hydrofoils and hovercraft

The drag force which slows down ships

Large sailing ships can still be seen. The many sails have to be carefully adjusted to make maximum use of the wind.

can be very much reduced by lifting as much as possible of the vessel above the water surface. Hydrofoils and hovercraft utilize this principle.

In a hovercraft, or air-cushion vehicle, air under pressure lifts the vessel above the surface it rides on. This surface can be land or water. Powerful fans drive the air through slots on the underside of the vessel. The hull is buoyant, enabling it to float when it is not supported by the air cushion. It is surrounded by a tough 'skirt', made of a tyre-like material, which contains the air cushion. Surface resistance is very much reduced and these vehicles are capable of travelling at high speeds.

When travelling at speed, the hydrofoil rises out of the water on its legs which act in a similar way to the aerofoils of aeroplanes.

Pressure

Pressure is the force acting on each square metre. This means that when a load is supported over a larger area, the pressure is less.

Using pressure

Tractors and bulldozers have large tyres or tracks. These reduce the pressure that they exert on the ground so that it does not make them sink in, even on soft ground. A car driven on to a ploughed field would probably sink in and get stuck, whereas a tractor could drive across it easily. When the same weight is carried over a smaller area, the pressure is increased.

Snow-shoes can be fitted to ordinary boots. The effect is to increase the area over which the weight is spread, and so reduce the pressure, since snow can only support small pressures.

In order to cut into something, the pressure must be increased until it is more than the strength of the surface can support. A knife has a very sharp edge, which means that the load on a knife blade pushing down is spread over a very small area indeed. This is why it cuts in. Razor-blades are even sharper than knives, and you can easily cut your-self without even knowing it. (**Don't try it!**)

Builders use pile-drivers to force piles (parts of foundations) into the ground. A large force (from a dropping weight) is applied to the top of the pile, which is quite narrow. The resulting large press-ure causes the pile to penetrate the

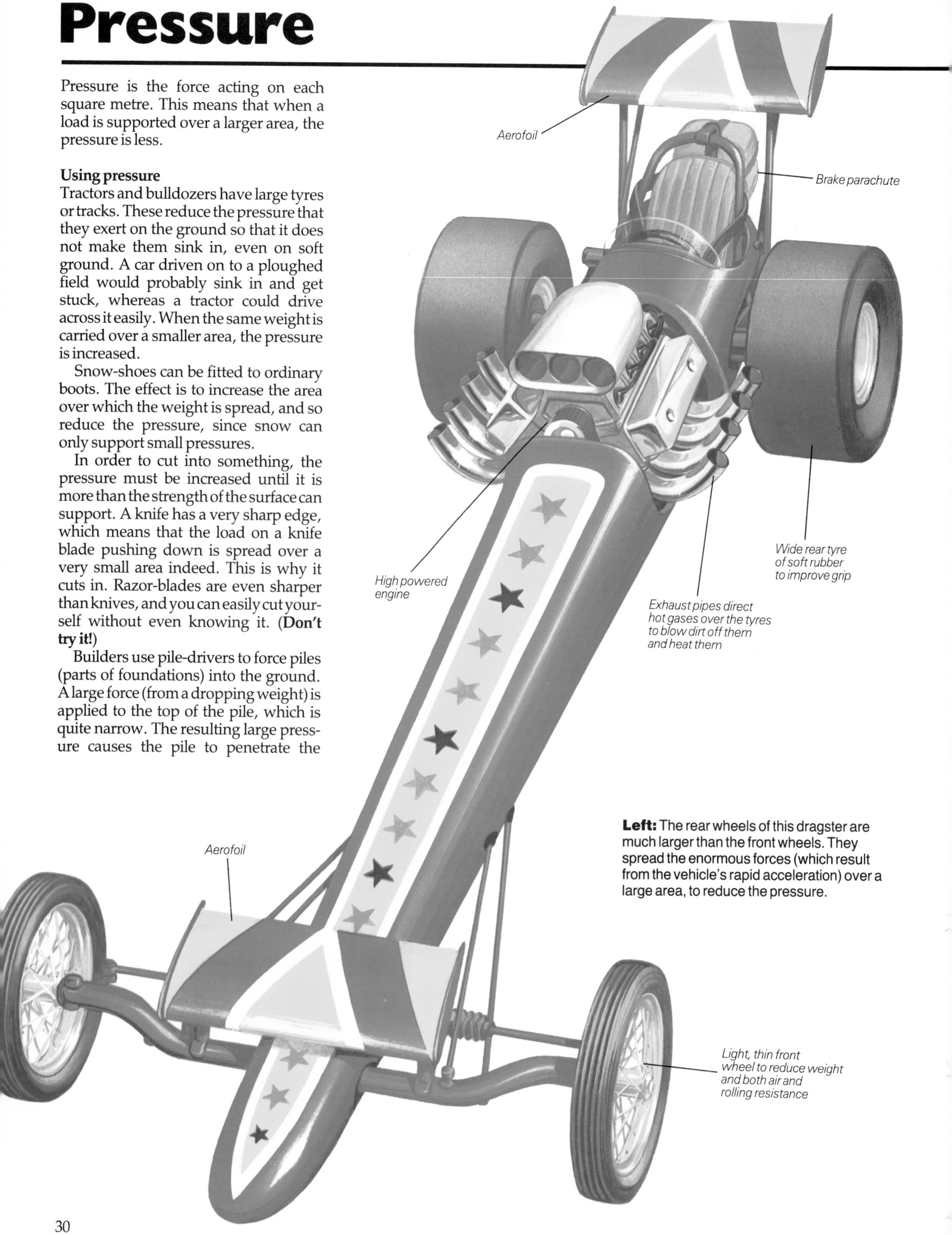

Aerofoil

Brake parachute

High powered engine

Exhaust pipes direct hot gases over the tyres to blow dirt off them and heat them

Wide rear tyre of soft rubber to improve grip

Aerofoil

Left: The rear wheels of this dragster are much larger than the front wheels. They spread the enormous forces (which result from the vehicle's rapid acceleration) over a large area, to reduce the pressure.

Light, thin front wheel to reduce weight and both air and rolling resistance ground. A garden fork works in a similar way. A force (due to the gardener's weight) is applied to the top of the fork, and the narrowness of the tines (prongs) gives a high enough pressure to cut into the ground.

Hydraulics and pneumatics

The examples given so far have concerned the pressure of solids. The study of pressure in gases is called pneumatics and in liquids, hydraulics. Unlike solids, liquids and gases can flow freely in all directions, so pressure can be transmitted in all directions. For example, water can be forced down a hosepipe, even if it is not straight. In the same way, pressure forces gas from the microscopic cavities in the North Sea bed along the twists and turns of the pipes to where it is used.

This property of transmission of pressure is very useful. The hydraulic brake system on cars allows the motion of the brake pedal to be sent to each wheel, even though they may be bouncing up and down as the car goes over bumps. Water is not used as the hydraulic fluid because it would cause corrosion, and would boil when heated by friction in the brakes. A special oil is used instead.

Hydraulic systems are used in aeroplanes to control the flap and rudder positions. Until recently, there was a network of pipes, in central London, used for transmitting water at high pressure to work machines. But the coming of electrical power distribution caused this system to be abandoned.

The actual force on a piston from a liquid under pressure depends on the area of the piston; the larger the area, the larger the force. The hydraulic press and hydraulic jack depend on this principle. In both cases, a force on a small piston causes a pressure in a liquid, which in turn acts on a large piston, which therefore exerts a much larger force. Hydraulic pressures are used to crush old cars, and hydraulic jacks are used in garages.

Pneumatic devices work in much the same way as hydraulic ones; many points on railways are operated by compressed air, which you can hear when they operate.

Right: Until recently, hydraulic power was used at Tower Bridge, London, to raise the bridge. The system was powered by steam engines, which pumped water at very high pressures along pipes to where it was needed.

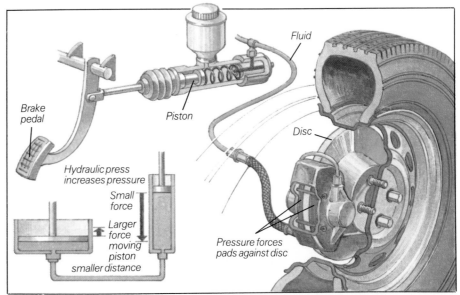

Above: The brake system of cars depends on the transmission of pressure by liquids. **Below:** Bulldozers have tracks so that their weight is spread over a larger area.

Aircraft

Left: These *Mirage* jets can travel at Mach 2.2, that is, more than twice the speed of sound. Air resistance makes its surface heat up so that it expands considerably!

One of the most complicated machines you could find is an aeroplane. In a large modern jet there are a great number of controls and dials for the pilot and crew to keep a check on.

Although the details of aircraft machinery are often very complicated, the principles of flight are fairly easy to understand.

Lift

One or more engines provides a forward force or thrust. The thrust from the engine accelerates the aeroplane forwards. As it moves faster, the air resistance, or drag, increases. When the thrust from the engine just balances the air resistance, the aircraft flies at a steady speed.

At the same time, the wing develops an upward lift force that depends on the speed of the plane. If the lift generated by the wing is greater than the weight of the plane, it will accelerate upwards.

It is easy to demonstrate the way a wing is lifted as the air passes over it. Hold a sheet of paper horizontal, near your mouth. Blow over the top of it; it should rise up and remain like that until you stop blowing. Because of differences in air speed above and below the wing, air pressure is reduced on the upper side, giving upward thrust.

Suppose a plane is flying at a steady height and at a steady speed. In this case, the upward lift is exactly balanced by the weight of the aircraft, and the thrust is exactly balanced by the air resistance. If the thrust from the engines is increased by more fuel being fed to them, the plane will accelerate until the drag increases to match the new thrust. However, as the plane accelerates, the lift from the wings increases and the plane will start to rise. In order to stay at the same height, the lift must not increase. Therefore, the wing is provided with ailerons (flaps) that can be adjusted so that the wing produces less lift. When you realize that the plane can also move left, right, up or down, and can twist in all directions (pitch, yaw and roll) you will appreciate why the pilot needs all those indicators and controls.

Control surfaces

The 'control surfaces' are the moveable parts of the aeroplane that can affect its flight. They include the ailerons, which operate either together to increase or decrease the lift from the wings, or in opposite directions to tilt the aircraft. Other control surfaces are the rudder, tail flaps (elevators), leading edge slats, spoilers and air brakes. These are either designed to change the direction of the plane, or to make it more efficient in flight. For instance, the elevators are spread out so that the plane can take off and land at low speeds. Because these extra flaps would cause more drag at cruising speed (and therefore give worse fuel consumption) they have to be folded back once the plane is in flight.

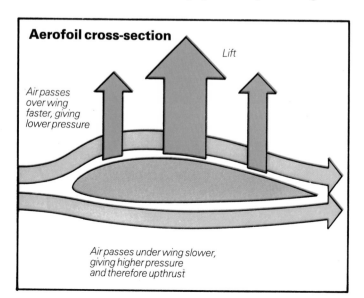

Aerofoil cross-section

Lift

Air passes over wing faster, giving lower pressure

Air passes under wing slower, giving higher pressure and therefore upthrust

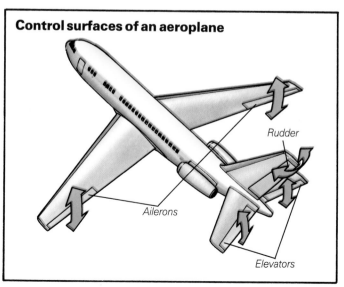

Control surfaces of an aeroplane

Rudder

Ailerons

Elevators

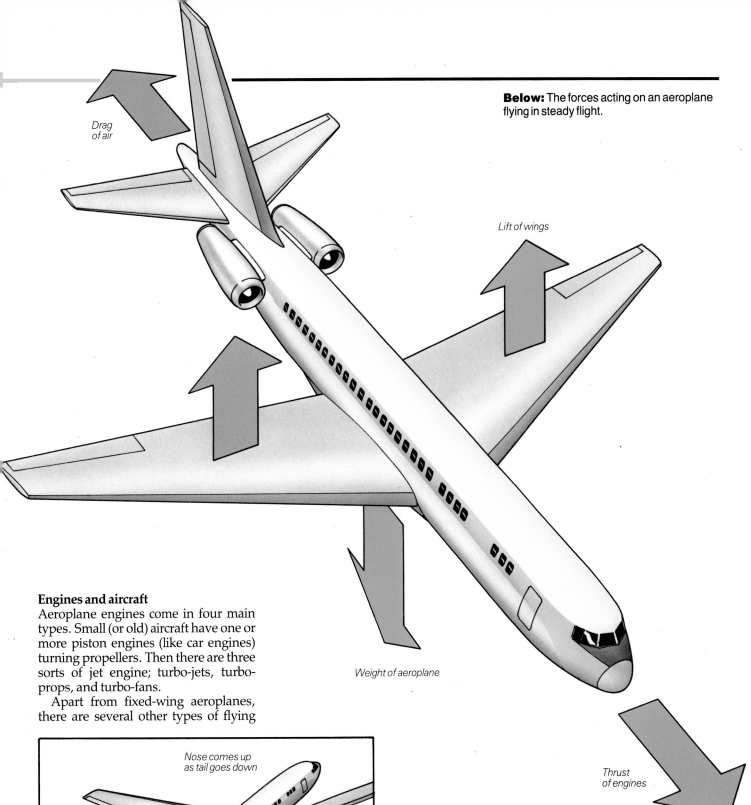

Drag
of air

Below: The forces acting on an aeroplane flying in steady flight.

Lift of wings

Engines and aircraft

Aeroplane engines come in four main types. Small (or old) aircraft have one or more piston engines (like car engines) turning propellers. Then there are three sorts of jet engine; turbo-jets, turbo-props, and turbo-fans.

Apart from fixed-wing aeroplanes, there are several other types of flying

Weight of aeroplane

Thrust
of engines

Nose comes up
as tail goes down

Elevators up

Airflow

Airflow across elevators
forces tail down

Left: When an aeroplane is climbing, the plane is tilted to provide more lift, and the engines are producing more thrust.

machines. The glider, which is un-powered, has to be towed either by a winch or an aeroplane, to make it move. Even so, by skilful use of air currents, gliders can stay aloft for considerable periods, and cover large distances.

The helicopter is an aircraft supported by a rotating wing. The helicopter is moved by tilting the rotor in the desired direction. The rotor blades are provided with variable pitch (tilt) controls, to cope with different conditions.

Rockets and satellites

Below: An example of Newton's Third Law. When the gun fires, the forward force on the shell is equal and in the opposite direction to the recoil due to the backwards thrust.

Shell

Position of gun after firing

Original position of gun

Below: The solid fuel in this firework rocket causes hot gas to be thrown out of the bottom very fast. The reaction drives the rocket upwards.

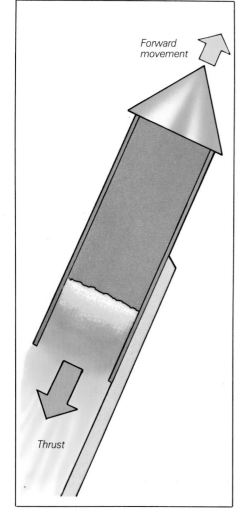

Forward movement

Thrust

Rockets and jet engines use the same principle for propulsion; that is, they both develop thrust by 'throwing away' large amounts of hot gases very fast. The principle that rockets use is Newton's Third Law, that is, 'Action and reaction are equal and opposite'.

Fuels

Jets use oxygen from the air to burn their fuel. Since rockets travel in space, they have to carry both fuel and oxidant. The oxidant may be oxygen, or it may be another chemical rich in oxygen.

Many different fuels and oxidants are used in rockets. But all rocket fuels must be able to give out a great deal of energy very quickly. Firework rockets use a solid fuel mixture that contains both fuel and oxidant. Most modern rockets use liquid fuels, but the NASA space shuttle uses two enormous solid fuel boosters which use a powdered mixture of aluminium perchlorate, aluminium oxide, and iron oxide mixed with rubber, to form a solid block. Each booster uses 560 tonnes of fuel at a rate of 4.7 tonnes per second and is burnt out in only 2 minutes, when it is discarded.

Liquid fuel rockets usually use liquid hydrogen as the fuel and liquid oxygen (LOX) as the oxidant. The hydrogen and oxygen are carried in liquid form because they take up less room than as gases. They are liquefied by a combination of high pressure and very low temperatures. Each of the three liquid-fuelled engines on the space shuttle uses 500 kilograms of fuel every second. The

advantage of liquid-fuelled rockets is that they can be controlled; once you have lit a solid-fuelled rocket you cannot turn it off! The disadvantage of liquid-fuelled rockets is that the pumps and piping which are needed have to withstand the extreme temperatures and vibration.

The space shuttle

Together, the three main engines and the two boosters fitted to the space shuttle give a thrust of more than 30 times the weight of the loaded shuttle. This accelerates the shuttle to a speed of 4800 kilometres per hour in 2 minutes, by which time it has reached a height of 45 kilometres above the Earth. Once the shuttle has reached a height of about 208 kilometres and a speed of 27,000 kilometres per hour, the engines are turned off and the shuttle orbits freely above the Earth's atmosphere.

Satellites

How can satellites orbit the Earth without falling to the ground, even though they are affected by the Earth's gravity? Newton discovered the explanation of this. Suppose a gun fires a shell – it will eventually hit the ground. If the gun was much more powerful, the shell would travel further before hitting the ground. But, remember, the Earth is round. If the shell were fired fast enough, it would fall around the Earth. This is what happens to satellites; they are not beyond the force of Earth's gravity, but they are falling round the Earth.

You may have heard of astronauts being 'weightless', or being in 'free fall': all this means is that they do not feel the Earth's gravity because they are falling. In the same way, you feel no weight when you dive off a spring board. Of course, because you are not in orbit, you soon hit the water!

The atmosphere extends a long way into space, becoming thinner further out. Satellites less than about 150 kilometres up slow down because of the air resistance, and eventually fall down. Because they are falling so fast, they burn up as they enter the thicker part of the atmosphere. Spacelab, a large American satellite, fell in this way. A Russian nuclear powered satellite fell in Canada, and may have caused pollution because of its waste products.

A satellite 35,900 kilometres above the ground orbits the Earth in exactly 24 hours. If the satellite is above the equator, it appears to be stationary there because it moves at the same rate as the Earth rotates. This is a geostationary or synchronous orbit and such orbits are frequently used for communications satellites. The advantage of these satellites is that moveable aerials are not needed. Telephone links between Europe and the USA have been by satellite since 1965.

Newton's ideas on satellites

Satellite orbits

1. Circular orbit
2. Elliptical orbit
3. Highly elliptical orbit
4. Escape

Above: A satellite system. Communications satellites can both receive and transmit telephone and television signals across the Earth. The areas receiving the transmitted signals are circled.

Boosters separate

External fuel tank separates. Shuttle goes into orbit

Satellite released

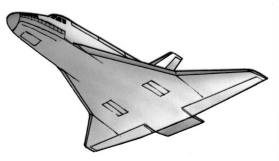
Re-entry into atmosphere

Launch

Above: The space shuttle takes off, with power provided by burning fuel from the external fuel tank and the boosters which are later jettisoned. After carrying out the space mission, the shuttle returns and glides to a landing without power.

Landing

Wind and water power

Machines are built with the purpose of helping people to do work. The simpler ones can be operated by muscle power, but there are many jobs which even with a machine are difficult or impossible to do using muscle power.

Water mills

People realized that grinding corn could be made easier by using something better than muscle power – and so water power was used to turn a large millstone which crushed the ears of wheat. The earliest kind had water wheels which rotated horizontally – like modern turbines. These can only be used with a fast flow of water.

The simplest kind of water wheels have large bladed, vertical wheels dipping into streams. They are called undershot wheels. In the overshot wheel, the water runs over the top of the wheel. This kind is two or three times more efficient than the undershot type but more expensive to build.

There are not many water mills left nowadays. They were very common until the industrial revolution when steam engines powered by coal replaced them. But coal has to be mined, and supplies will eventually run out. Running water will always be found wherever and whenever rainwater collects on land.

Hydro-electric power

Water power is used very successfully on land to generate electricity. This is done by hydro-electric power (HEP) stations. Water flowing at high speed turns

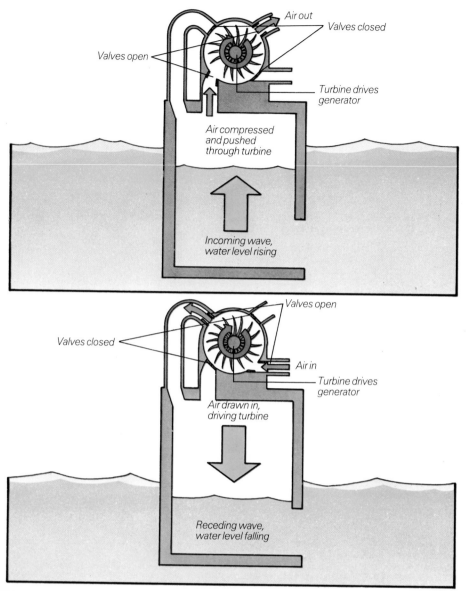

Air out
Valves closed
Valves open
Turbine drives generator
Air compressed and pushed through turbine
Incoming wave, water level rising

Valves open
Valves closed
Air in
Turbine drives generator
Air drawn in, driving turbine
Receding wave, water level falling

Above: Wave power generator. This device is used to light small navigational buoys in Japan. The rising waves compress the air in the tank. This turns the turbine which causes generation of electricity for the light. When the wave falls, air rushes in, again turning the turbine.

Left: An overshot water mill.

a turbine which drives a generator producing electricity.

The water is usually piped from a dammed reservoir at a higher level. Large schemes need rainy, mountainous regions, smaller schemes can use fast flowing rivers. At times when less water is being used (for example, at night), the excess power can be used to pump the water back up to the reservoir. Thus the energy is stored until it is needed. This is called a 'pumped storage' scheme.

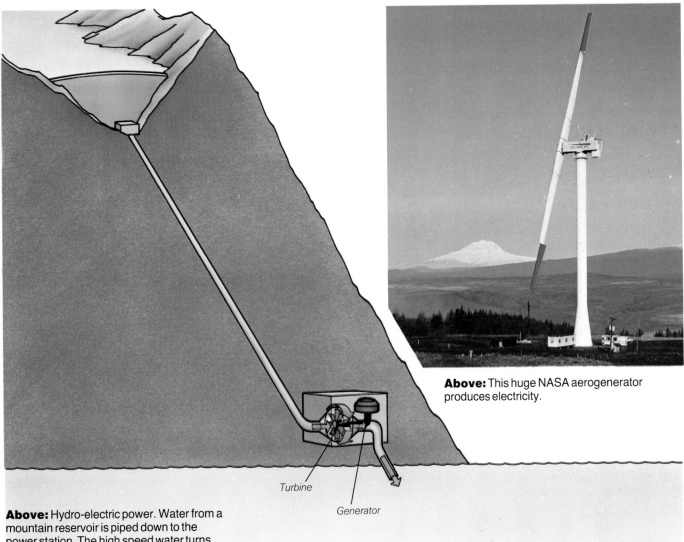

Above: Hydro-electric power. Water from a mountain reservoir is piped down to the power station. The high speed water turns the turbine. The spinning turbine runs an electric generator.

Turbine

Generator

Above: This huge NASA aerogenerator produces electricity.

Power from the sea

The seas and oceans obviously contain a lot of energy but both the possibilities and problems of using it are enormous.

The problems are that waves come and go according to the weather, and the best waves are out at sea. Not much energy is needed there, though the Japanese have successfully powered electrically lighted signal buoys by wave power. One possibility is for machinery out at sea to generate electricity which is brought ashore by cables. The machines would have to be very large to be efficient, and because of the expense, such a scheme has not yet been built, but different designs are being tested.

Another way to get water power from the sea is to use the tidal movement, since twice a day the water level rises by several metres. When the level falls again, as the tide goes out, it could be made to turn turbines and generate electricity. The only place where this is used at present is in the dammed Rance estuary near St. Malo in northern France. Only certain estuaries are suitable and it is expensive to build the huge dams (housing the turbines) which are required. People are also worried about the possible effects on the environment of interfering with the normal tidal flow.

Windmills

It is more difficult to get much energy from moving air than from water. Winds are more variable, and don't contain as much energy. Nevertheless windmills are common throughout the world. They are usually used for driving simple machines such as mills for grinding, or for pumping water for drainage or irrigation. Traditional windmills have a horizontal axle with several vertical sails attached and these are turned by the wind. The sails can usually be turned to face the wind.

One of the recent windmill designs is the Savonius type which uses split oil drums for sails. These can be very cheaply made but do not give much power. At the other extreme there are the large, powerful aerogenerators – NASA has developed one with a single vane which looks like an aircraft propeller shaft.

For windmills to operate efficiently they need sites that have average wind speeds of over 30 kilometres per hour. In Britain only about 30 such sites have been identified. Alternatively, windmills could be placed out at sea, but this is more expensive. There may be an increase in small-scale use of wind power, but it is not expected that large scale wind machines could supply more than about one per cent of our energy needs.

Generation of electricity

Electricity is very important in our modern world. We find ourselves in chaos when there is a power failure. We are without light, television, power for electrical appliances and so perhaps have no heating. The reason we use so much electricity is because it provides power very conveniently—it is the power source for many machines.

The electric motor

It is easy to show how electricity can make something move. If a strong magnet is brought near to a copper wire carrying an electrical current (from a car battery, for example), the wire will move. This is because there is an electromagnetic force on the wire as it carries a current in a magnetic field. But such a fixed wire will not be able to move very far.

An electric motor allows this motion to continue, by reversing the current in the coil so that it changes its direction of movement, and when connected to a power supply the motor rotates continuously. The magnetic field in a motor is usually provided by coils of wire, called field coils, which form an electromagnet. (Fixed steel magnets may be used instead.) These surround the moving coils of wire, called the armature, which carry the currents. Several coils of wire are usually used to give a greater turning effect. Rubbing against the end

of the armature are carbon conductors or brushes which connect it with the power supply. This connection is through the brushes and the end-piece, or commutator, and it is this which changes the direction of the current. The movement of the armature turns the axle, which is connected to belts and pulleys or gears, to provide motion. There are many different designs of motor in use but the principle of how they work is the same.

Anything which moves continuously when attached to a supply of electric current must contain a motor. Take a look around your home to see just how many such objects there are.

Electricity supplies

Electric motors can run on very small currents from cells of batteries, but these are only suitable for toys. To do large amounts of work, a large supply of electricity is needed. This comes from the mains supply at 240 volts, or perhaps at a lower voltage through a transformer (also sometimes called a power supply). From the sockets in our homes, the supply system leads eventually to a power station where the electricity is generated.

The electric generator, sometimes called a dynamo or alternator, works on the same principles as the electric motor. It can be thought of as a motor with its functions round the other way. It, too,

consists of armature coils to carry current, field coils to cause a magnetic effect, and the commutator/brush arrangement for electrical contact with the rotating armature. (There are motor/generators available which can be either.) What happens in a generator is that the armature is rotated in the magnetic field and this generates the current – the reverse action to that in the motor. In fact, it does not matter whether it is the coil or the magnet which rotates in the generator – electricity is produced in both cases. The moving part is called the rotor and the fixed part is the stator.

Both the motor and the generator were invented by Michael Faraday, just over 100 years ago. But it was not until several years after he invented the motor that he discovered how to generate electricity. He quickly went on, however, to invent the transformer as well! Faraday's discoveries have revolutionized industrial and domestic life.

Recently, the Electricity Council in Britain celebrated the centenary of the first public electricity supply. Now, everywhere in Britain is connected by one supply because the National Grid of wires links all the country's power stations. This means that changes can be made to power stations, or old ones replaced or existing ones serviced, without cutting off the electricity supply.

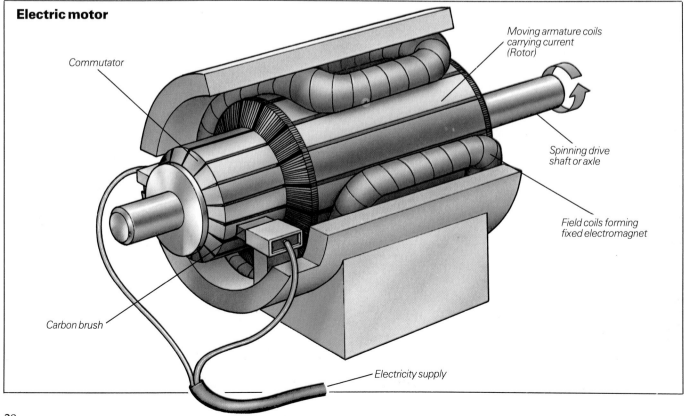

Electric motor

Commutator

Moving armature coils carrying current (Rotor)

Spinning drive shaft or axle

Field coils forming fixed electromagnet

Carbon brush

Electricity supply

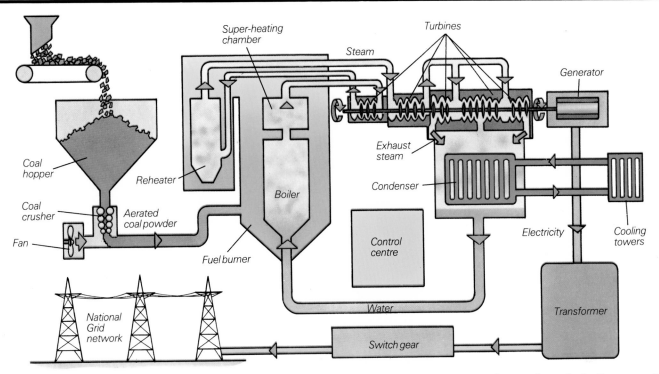

Super-heating chamber

Steam

Turbines

Generator

Coal hopper

Reheater

Boiler

Exhaust steam

Condenser

Cooling towers

Coal crusher

Aerated coal powder

Fan

Fuel burner

Control centre

Electricity

National Grid network

Water

Transformer

Switch gear

Power stations

Power stations are complicated and expensive to build. There are different types according to the fuel they use. Fuel can be coal or oil (fossil fuels) or nuclear power. Oil is sprayed into, or ground-up coal is blown into, the heat generator where the fuel burns. Nuclear fuel undergoes a controlled nuclear reaction

Below: An external view of three of the five turbines in a power station. The first two turbines produce 58% of the power, and exhaust steam from the second turbine drives the three other turbines.

Above: The workings of a power station. Huge steam-driven generators produce our electricity supply. These generators are amongst the most powerful machines we use.

which releases a lot of heat. In every case, resulting waste products have to be removed.

The heat generator usually boils water to produce steam at high pressure. This drives large turbines and their rotary motion turns the generator coils to produce electricity. A lot of heat is lost in this whole process. Some of this can be re-

covered to pre-heat the boiler water, but most is lost through the cooling towers that are an obvious feature of power stations. These serve to cool the water from the condenser which converts exhaust steam into water for the boiler.

The electricity produced then travels through transformers and switch gear to give the correct voltage for the National Grid power supply.

Below: The generator. Its rotor is an electromagnet connected to the turbine drive shaft and it produces current in the stator.

Turbines

Machines at work

With ready supplies of electricity and machines, we are no longer restricted by the limits of human abilities in supplying our needs. In earlier times, when people had to rely on their own strength in producing food, most people had to work on the land. Growing crops and raising animals to produce food was a time-consuming activity. In many parts of the world it still is. But in the western world, few people now work on the land although more food is produced than ever before.

Farm machinery

The tractor has become the farmer's constant companion. Fifty years ago this would have been the horse or some other work animal. A tractor is an all-purpose machine – it can be fitted with attachments to do most jobs connected with crop growing. It can pull a plough or harrow, dig, mow, bale or spray crops.

Many machines have been designed for gathering crops; for example, potato lifters, fruit tree shakers, and huge cotton boll pickers. The most widely used of all harvesting machines is the combine harvester for cereal crops. It does the work previously done by several people and in less time. The harvester first cuts the crops (reaping). It then separates the grain from the stalks (threshing). The outer cover of the grain (chaff) is blown away, and the stalks of straw are dropped out for baling. The grain (usually wheat, oats or barley) is dropped separately into a trailer alongside.

Factories

Most people in this country now live in large towns and many people work there in factories making anything from nails to television sets, from clothes to cars.

Factories in Britain really began with the discovery, in the nineteenth-century industrial revolution, of how to harness steam power. One of the first industries to be transformed by steam power was the making of cloth from cotton. As a result, although no cotton was grown in Britain, cotton cloth was the country's main export for 150 years, until 1938. The main export then became machinery. Britain is sometimes called the first industrialized nation for this reason.

Machines have many advantages in manufacturing. Making cloth, like harvesting wheat, involves many repetitive processes. Machines can do such work quickly and reliably. This means the goods are cheaper to the customer and make more profit for the manufacturer. A T-shirt costs about the same as the pay for one hour's work but takes much less than an hour to make. If it were made by hand from the cotton fibres, it would take several days!

Safety

Machines also make work a lot safer. Coal and other kinds of mining today use huge cutting and drilling machines instead of people putting themselves at risk. Imagine being a miner digging far underground, as many children had to do in the past. The chemical and metal manufacturing industries involve very high temperatures, and corrosive and poisonous substances. Without machines, some of this work would be impossible.

Workshops are needed by all industries to make tools and repair equipment. This means working with metal, plastic and wood. Lathes, drills, milling and cutting machines similar to those in school workshops are used in large numbers. This provides work for skilled operators, but it also means that, particularly on the production lines of large industries, people have been replaced by machines. This leads to people having difficulties in finding jobs.

In less industrially developed countries, such as Asia and Africa, there is not enough money to pay for advanced technology. Many people still work as labourers. But simple machines, like those described earlier in the book, may be put to good use.

Left: A nineteenth century factory. A steam engine drove the rotating shaft which is just below the roof and belts ran from this to operate the many machines below.

Below: Production lines, like this one in a light bulb factory, speed up work and require few operators.

Automatic machines

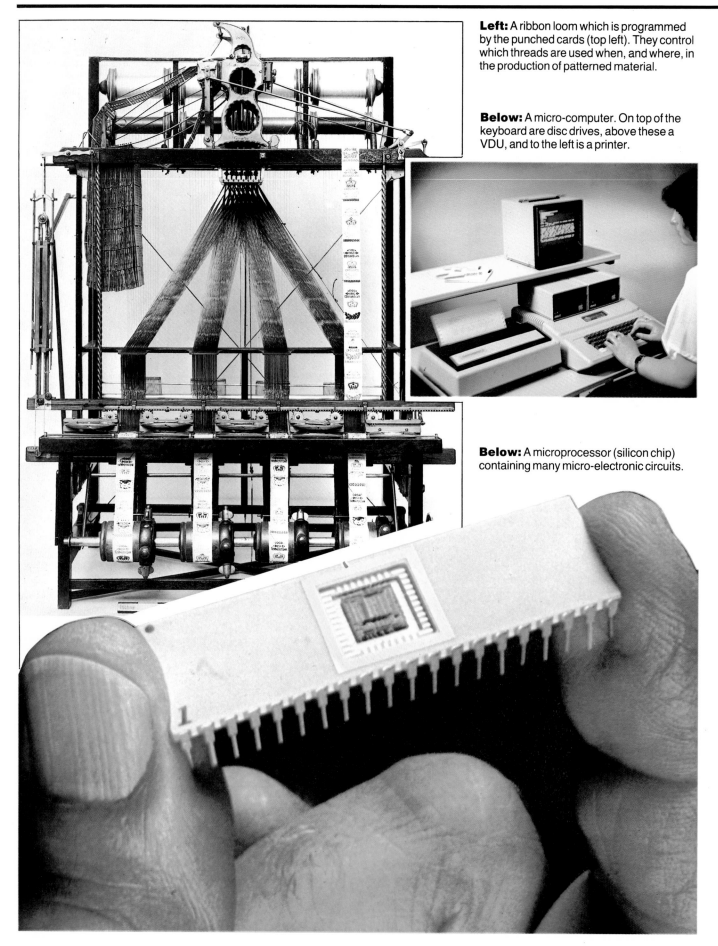

Left: A ribbon loom which is programmed by the punched cards (top left). They control which threads are used when, and where, in the production of patterned material.

Below: A micro-computer. On top of the keyboard are disc drives, above these a VDU, and to the left is a printer.

Below: A microprocessor (silicon chip) containing many micro-electronic circuits.

All machines need both energy to drive them and a control system. Control systems do not always need a person to operate them – the first automatic control was probably James Watt's fly-ball governor. He invented this in 1788, to control the speed of a steam engine. The balls are attached to a vertical axle which rotates faster when the engine goes faster. The balls then move outwards and upwards, away from the axle as it spins, closing down a valve as they do so and reducing the steam flow into the engine. Therefore the engine goes slower and the balls fall again.

The governor is a closed loop system. The fly-ball automatically slows the engine when it increases speed. There is negative feedback to make the conditions stable. Positive feedback causes instability, as when a pop group's microphones pick up the loudspeaker output and a screaming sound is heard.

Another important early invention was the punched card. (Similar cards are still used now to program computers.) Joseph Jacquard used this to control the weaving of silk cloth from 1801. The punched card is part of an open loop system. Control is put in at the beginning by positioning the holes. This is the program and the loom follows it exactly until the process is finished.

Electronics

The science of electronics has made it possible to develop many kinds of control. Light, heat and other signals can be converted into electrical currents and processed by complex, tiny, integrated circuits. These can then control much larger currents which operate machines. Systems of this kind are called servomechanisms. They are used in, for example, autopilots, machine tools such as lathes, and radar controlled guns. Robots are complete machines which are controlled in this way – they usually contain several servomechanisms.

Programmed machines

Robots are programmed in their actions. The program can be written before any robot activity occurs but can include the potential for additions by the robot, by feedback loops. This means that the robot can receive instructions to vary its actions as required, so it can seem that the robot has a mind.

This 'intelligent' part of the robot is a

Right: Robot welding machines in a car factory.

computer. Although there are many kinds, they all have the same basic parts. The central processor unit (CPU) is a high speed calculator with a clock and logic system to organize its operations. There are memories to store information and programs. These can be fixed in the computer and added to by means of magnetic discs or tapes.

Communication with a computer is by an input device. This is usually a keyboard on which to type messages. The computer communicates through an output device. This is often a visual display unit (VDU), for example, a television screen. A loudspeaker enables a computer to speak; a servomechanism enables it to move things.

Above: Artist's impression of robots building a space station.

Although computers are very good at doing a lot of things, they are actually no more than a complicated collection of switches. They work on a binary logic system; that is, each part is in one of two states – on or off. Computers are able to do things in a fraction of the time it takes us. To send a spacecraft to the Moon needs so many calculations, it would be impossible without computers. To build a space station – perhaps to obtain energy to drive our machines – will depend even more on computers. But will one be built? Only people can make that decision.

A-Z Glossary

Acceleration The rate of increase of speed. An object is accelerating if its speed is changing.

Aerofoil A body shaped so as to produce lift.

Alternator A machine which converts mechanical energy into electrical energy. An alternator produces alternating current (AC); a dynamo produces direct current (DC).

Aqua-planing A dangerous condition when the tyres of a car or lorry are pushed away from the road surface by a film of water, reducing friction to almost zero.

Armature A set of copper coils on an axle, in an alternator or electric motor.

Ball bearing A device for greatly reducing friction on rotating shafts. The shaft is supported by steel balls held in a cage.

Bolt and nut A pair of objects with matching screw threads. When either the bolt or the nut is turned, it moves with great force. Often used as fixing devices.

Cam A small, shaped piece of metal that turns on a rod to open a valve.

Carburettor A part of the internal combustion engine where the fuel is vaporized and mixed with air.

Cell A chemical source of electricity; sometimes it is called a battery.

Centre of buoyancy The position in a floating object where the upthrust appears to act.

Centre of gravity The position in an object where the weight appears to act; the balance point.

Centripetal force A force which acts towards a central point, as, for example, does the Earth's gravity.

Closed loop system A self-adjusting system which keeps conditions stable and is controlled by negative feedback.

Communications satellite Usually, a satellite in geo-stationary orbit, ie. one that appears to remain in the same place above the Earth as it rotates. It does not require steerable aerials on the Earth.

Counterbalance A weight which balances an equal weight working in opposition.

CPU Central processing unit – the part of a computer which processes information.

Crankshaft The part of an engine driven by the piston which makes it rotate.

Displacement The amount of water pushed out of the way by a ship when it floats.

Drag The force of air or water resistance acting against the motion of a body.

Dynamo See Alternator

Effort The force which is doing work on an object.

Engine A device for changing fuel energy to mechanical energy.

Feedback The return of part of the output of a system to the input. Negative feedback causes self-adjustment of the system and therefore stability: positive feedback causes instability.

Force A push or a pull, it causes objects to change their motion. It is a vector quantity, with a particular direction and forces must be combined with special vector rules.

Friction A force that tends to slow down relative motion. Heat is generated when sliding takes place.

Fulcrum See Pivot.

Gears Interlocking toothed wheels for the transmission of power.

Generator See Alternator.

Gravity The attractive force exerted by one body on another. All bodies which have mass have this property.

Heat That which increases the internal energy of a body. This causes a change in temperature, volume or state of the body.

Horsepower A unit of power equal to 746 watts. It is also equivalent to the power used in raising a 250 kilogram weight 30 centimetres in 1 second.

Hovercraft A ground vehicle that is supported by a cushion of air to reduce friction. As well as travelling on the land, hovercraft can travel on the sea or swampy terrain.

Hydraulics The study of pressure and flow in liquids. A hydraulic jack uses oil under high pressure to lift large objects more easily.

Hydro-electric power Electricity produced by using the kinetic energy of water.

Hydrofoil A vessel which skims the surface of the water and the shaped pieces on the bottom of the vessel which act like 'water wings' to give it lift.

Inclined plane A wedge used to raise a load more easily. The load moves, not the wedge.

Inertia The property of matter which makes it continue in a state of rest or in a state of motion. Force is needed to overcome inertia.

Insulator A substance which is a poor conductor of heat and/or electricity.

Internal combustion engine An engine in which the fuel is burnt within the working cylinder.

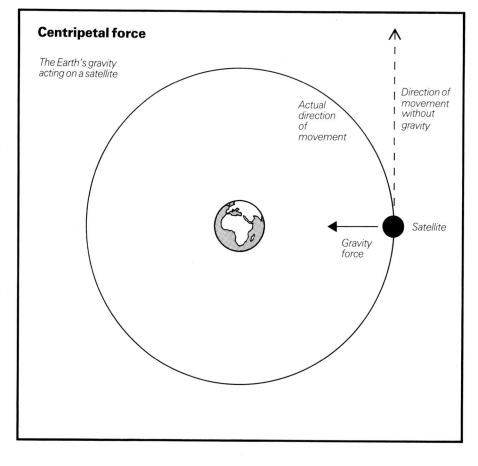

Centripetal force

The Earth's gravity acting on a satellite

Actual direction of movement

Direction of movement without gravity

Gravity force

Satellite

Jack A device, often containing a screw thread, for lifting heavy objects.

Joule The international unit of measurement of energy, work and heat. One joule equals the work done when a force of 1 newton moves a body 1 metre.

Kinetic energy The mechanical energy possessed by a body due to its motion. It may be calculated from the formula:
kinetic energy = ½ mass x (velocity)^2

Lever A simple machine comprising an arm and a pivot.

Load The object which is being lifted or accelerated.

Lubrication A method of reducing friction by putting a liquid between two sliding surfaces.

Magnetic field The region around a magnet or electromagnet where there is a magnetic force.

Mass The quantity of matter a body contains. Mass is measured in kilograms (and often incorrectly called weight). The mass of a body does not change if, for example, it is moved to the Moon where the force of gravity is less. (See Weight.)

Momentum A measure of a body's motion. It can be calculated from the product of the body's mass and velocity.

NASA National Aeronautics and Space Administration (of the USA).

Newton A unit of force. One newton is the force which acts on a mass of 1 kilogram to produce an acceleration of 1 metre per second per second.

Nuclear Involving the inner part of an atom, the nucleus. Changes in the nucleus can release a lot of energy which can be used in the generation of power.

Open loop system A control system which operates by feeding in instructions at the beginning of a process and they are followed to complete the process.

Pile This is a steel or concrete bar which is driven into the ground to form part of the foundations for a building.

Piston The part of an engine that moves up and down in the cylinder after the explosion of the fuel.

Pivot The place at which a lever swivels. A lever cannot work without a pivot.

Pneumatics The study of pressure and flow in gases. A pneumatic drill is operated by compressed air; a pneumatic tyre contains compressed air.

Potential energy The mechanical energy possessed by a body due to its position. For example, water in a reservoir above a hydro-electric power station has high potential energy which is converted to work when it drives the turbines of the power station.

Power The rate of doing work.

Program A set of instructions telling a computer what to do.

Pressure The force acting on each square metre. To calculate pressure, divide the force (in newtons) by the area (in square metres). It is measured in newtons per square metre (N/m^2).

Pulley A grooved wheel carrying a string or rope which turns the wheel.

Radio-active Giving out radiations which can pass through the body and may cause death or illness.

Reciprocating Moving backwards and forwards, like the piston in a car engine.

Roller A device for replacing sliding friction with rolling friction.

Roller bearing Similar to a ball bearing but using rotating steel cylinders.

Satellite An object in orbit around the Earth or another planet. The Moon is a satellite of the Earth.

Screw A spiral path along a cylinder. A screw is like an inclined plane wrapped up. It can take a rotating force and change it into a much larger straight-line force.

Spacelab A large manned NASA satellite that re-entered the Earth's atmosphere and was burnt up. No astronauts were aboard at the time!

Speed The average rate at which an object is travelling. Measured in metres per second (m/s), kilometres per hour (km/hr), or miles per hour (mph).

Spring balance An instrument for measuring weight or force which does so by the elasticity of a spiral spring.

Suspension system The part of a vehicle which connects the wheels to the body, so that the ride is comfortable.

Transformer An arrangement of copper coils with an iron centre, used to change the voltage of an electric current.

Transmission system The part of a vehicle which carries the movement of the engine to the wheels.

Triangle of forces If three forces acting at a point can be represented in size and direction by the sides of a triangle, and the directions follow round the triangle, the forces are in equilibrium. If only two forces are known then the third can be measured if the triangle is drawn to scale. The example

below enables measurement of the force acting vertically on the suspended weight in the diagram on page 7. The upper side of the triangle here is drawn in a position parallel to the position of the larger force exerted by the left-hand string and spring balance. Thus a triangle of forces combines two forces to give the resultant force.

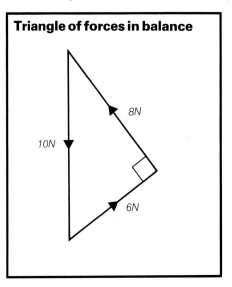

Triangle of forces in balance

10N 8N 6N

Turbine A machine for generating rotary power. Blades fixed to a rotating shaft are turned by air, steam, gas or water.

Upthrust The upward force produced when an object is immersed in a liquid or gas.

Valve A mechanism which controls the flow of gas or liquid down a tube.

VDU A visual display unit, similar to a television screen, which is used to display information from a computer. It is controlled by a keyboard.

Vector A physical quantity which has direction – for example, force, momentum.

Velocity The rate of motion in a particular direction.

Watt The international unit of measurement of power. One watt equals 1 joule per second.

Wedge A simple machine cosisting of a triangular shaped block, that can be forced into a gap, giving a greatly increased force.

Weight The force of gravity acting upon a body. It is measured in Newtons. (See Mass.)

Weightlessness A condition when an object is in free fall. It does not mean that the object is completely out of the Earth's gravitational field.

Work Production of an effect by exertion of a force. Energy is needed to do work and work is measured in joules.

Reference

Books to read

Machines and Energy by Eric J. Barker and W.F. Millard (Arco, 1972)
Wind Power, Water Power, Solar Power and *Electric Power* by Ed Catherall (Silver Burdett, 1982)
Man and Space by Arthur C. Clarke (Time-Life, 1982)
The Space Shuttle by George S. Fichter (Franklin Watts, 1981)
The Simple Facts of Simple Machines by Elizabeth James and Carol Barkin (Lothrop, Lee, and Shepherd, 1975)
Machines by Robert O'Brien (Time-Life)

The history of machines and mechanics

THE STONE AGE: BEFORE 8000BC
The hand axe This was probably the first technological item ever made. It is just a piece of stone, sharpened by chipping with other stones. It was hand-held for cutting. Relics of axes have been found all over the world and date from about one million years ago.
Fire This was used in preparing wooden and stone tools.

THE BRONZE AGE: 8000-2000BC
Metal working Throughout this period, the ability to work copper, and later to extract it from impurities, was developed. Bronze is an alloy of copper and tin which makes a much stronger material, and this was widely used at this time.
(About 3500BC) The wheel The first one was the potter's wheel used in Mesopotamia (now Iraq) by the Sumerians. It had been developed for transport, as a cart wheel, by about 3000BC.
(About 2500BC) The lever, inclined plane, and wedge These simple machines were used by the Egyptians in the construction of pyramids. The lever was also used in the weighing balance and in the shadoof for raising water.
(About 2000BC) The plough Ox-drawn plough-shares were used in Mesopotamia and Crete.
(About 2000BC) The sailing ship There was regular travel up the Nile in simple square-rigged ships, sailing only before the wind.

THE IRON AGE: 2000-500BC
(About 1500BC) Iron smelting Wrought iron was produced in Mesopotamia by the Hittites and used for tools such as tongs, and weapons such as swords.
(About 1500BC) The chariot Fast, horse-drawn vehicles were developed in China and Mesopotamia.
(About 800BC) The pulley The simple one-wheel pulley was invented.
(About 600BC) Siege machines Large towers and battering rams on wheels were in use.

GREEK AND ROMAN TIMES: 500BC-500AD
(287-212BC) Archimedes The discoverer of the law of buoyancy (Archimedes' Principle), and the first person to study basic mechanics, he worked out the 'law of the lever' and used this in many machines, especially in the defence of his home town, Syracuse, against the Romans. He also invented the Archimedean screw for raising water for irrigation.
(About 250BC) The piston and valve These essential parts of today's car engines were used in a type of music organ by the Greek, Ctesibus of Alexandria.
(About 70BC) The water wheel The first one was used to grind corn in Rome.
(About 60BC) Hero This Greek from Alexandria made many mechanical inventions. These included the first steam engine, and numerous automatic machines operated by falling weights, water and steam. One was the first vending machine, a coin operated dispenser of holy water!

THE MIDDLE AGES: 500-1500AD
(About 1000AD) The wheelbarrow A large-wheeled version for earth moving was invented in China.
(About 1000AD) Paper This was invented by an official at the Emperor of China's court.
(About 1000AD) Gunpowder This was invented in China for use in fireworks. They were developed into rocket-driven weapons.
(800-1200AD) Navigation The science of navigation developed greatly under the Arab civilizations of this period. The astrolabe, for astronomical observations, became an essential aid for sailors in finding their directions by the stars. The dhow sailing ship was developed for exploration and trade purposes.

THE RENAISSANCE PERIOD: 1400-1700AD
(1400-1456) The printing press Type was cast in metal in Korea in the early 1400s. Johannes Gutenburg produced the first printed Bible in Mainz, Germany, in 1456, using a converted wine press!
(1452-1519) Leonardo da Vinci An Italian genius in things artistic and mechanical, he designed many machines that were not built until long after he was dead. The weapons he designed included a tank-like armoured vehicle. He also pioneered the study of aerodynamics with designs for gliders, a helicopter, and a parachute. His other designs included spinning and weaving machines, pumps, a hydraulic jack, a gas turbine, a primitive bicycle and an automatically rotating spit.
(1564-1642) Galileo Galilei An Italian whose work on falling objects disproved the ideas of Aristotle which had been the basis of mechanics for nearly 2000 years. He also discovered the principle of the pendulum and built one of the first telescopes. His observations on the Solar System and his

theories of motion established the ideas proposed by Copernicus (that the Earth is not the centre of the Universe) and made him very unpopular with the Church of that time.
(1642-1727) Isaac Newton He developed the science of mechanics on foundations laid by Galileo, producing laws which can be used to predict the movement of everything from atoms to planets. He discovered the law of gravitation and made important discoveries about the nature of light.

THE EIGHTEENTH AND NINETEENTH CENTURIES
(1698) The steam engine It developed in England, from the Savery pump (1698) to the Newcomen engine (1712) used in the Cornish mines. By 1775, Boulton and Watt's improved engines had laid the basis of Britain's industrial revolution.
(1733-1792) Textile machines The first automated industry in the world, using first water power and then steam power, was the cotton industry in Britain. Important inventions include: John Kay's flying shuttle (1739); John Hargreaves' spinning jenny (1764); Richard Arkwright's spinning frame (1771); Samuel Crompton's spinning mule (1779); and Eli Whitney's cotton gin in the USA (1792).
(1775-1800) Precision tools These were developed to make the new machinery and included cylinder boring lathes invented by John Wilkinson (1775) and screw cutting lathes invented by Henry Maudsley (1800).
(1800) The electric battery The voltaic pile was built by A. Volta of Italy to produce electric current.
(1804-1829) The steam locomotive The first one was built by Richard Trevithick; by 1829, George Stephenson had built two public railways, and the fastest locomotive, the 'Rocket'.
(1821) The electric motor Michael Faraday first produced continuous rotation of a magnet around a current-carrying wire.
(1822) The calculator Charles Babbage built the first (large, mechanical) calculating machine.
(1831) The electric generator Michael Faraday discovered electromagnetic induction.
(1840) Energy James Joule established that heat and mechanical work are aspects of the same thing – which later became termed 'energy'.
(1840-1850) Thermodynamics The science of heat flow, particularly important in both steam and internal combustion engines, was established in important theories from W. Thomson (Lord Kelvin) and R. Clausius.
(1845) Steamships Isambard Brunel designed the *Great Britain*.
(1846) The sewing machine It was invented by E. Howe in the USA, and manufactured by Isaac Singer.
(1885) The petrol-driven car In

Launch of the space shuttle.

Germany, the predecessors of the modern car were first built independently by Karl Benz and Gottlieb Daimler. Before this there had been a variety of steam and electric road vehicles built.

THE TWENTIETH CENTURY
(1900) The airship Count von Zeppelin launched the first petrol driven, rigid airship which was soon running a passenger and mail service.

(1903) The aeroplane The Wright brothers designed and flew (briefly!) *Flyer 1* in the USA.

(1934) The hydrofoil The first design was built by V. Grunberg; the modern hydrofoil was developed by 1964.

(1942) The nuclear reactor One was first built in Chicago by a team of physicists under E. Fermi.

(1948) The transistor This was invented in the USA by Shockley Bardeen and Brattain.

(1959) The hovercraft The first commercial vehicle came into service in the English Channel.

(1961) Space flight Yuri Gagarin of the USSR made the first flight around the Earth in Vostok 1.

(1969) The Moon landing Neil Armstrong and Edwin Aldrin of the USA landed on the Moon from Apollo 11.

Acknowledgements

Photographs
Aldus Archive: 22R, 29B
All-Sport/Tony Duffy: 8, 9
British Railways Board: 27L & R
British Tourist Authority: 29T
CEGB: 39
Colorsport: 19
Dunlop: 22L
Mary Evans Picture Library: 21
Ferranti Electronics Ltd.: 42BL
Fiat Auto UK: 43B
J. Freeman: 12
Thor Heyerdahl: 10
Mansell Collection: 41C
MARS/Avions Marcel Dassault-Breguet Aviation: 32
Microsense Computers Ltd.: 42T
NASA: 37, 43T, 47
Raleigh Ltd.: 17
Science Museum: 42BR
SNCF: 26
Spectrum Colour Library: 6
Stephens Belting Co.Ltd.: 14
Transport and Road Research Laboratory: 25
Volvo Concessionaires Ltd.: 24-5
ZEFA: cover, contents page, 5T & B, 16, 31, 36, 41T & B
Artists
David Eaton; Donald Harley, Jeremy Gower and R.A. Sherrington by courtesy of B.L. Kearley Ltd.

Index